Studies in Systems, Decision and Control

Volume 139

Series editor

Janusz Kacprzyk, Polish Academy of Sciences, Warsaw, Poland
e-mail: kacprzyk@ibspan.waw.pl

The series "Studies in Systems, Decision and Control" (SSDC) covers both new developments and advances, as well as the state of the art, in the various areas of broadly perceived systems, decision making and control- quickly, up to date and with a high quality. The intent is to cover the theory, applications, and perspectives on the state of the art and future developments relevant to systems, decision making, control, complex processes and related areas, as embedded in the fields of engineering, computer science, physics, economics, social and life sciences, as well as the paradigms and methodologies behind them. The series contains monographs, textbooks, lecture notes and edited volumes in systems, decision making and control spanning the areas of Cyber-Physical Systems, Autonomous Systems, Sensor Networks, Control Systems, Energy Systems, Automotive Systems, Biological Systems, Vehicular Networking and Connected Vehicles, Aerospace Systems, Automation, Manufacturing, Smart Grids, Nonlinear Systems, Power Systems, Robotics, Social Systems, Economic Systems and other. Of particular value to both the contributors and the readership are the short publication timeframe and the world-wide distribution and exposure which enable both a wide and rapid dissemination of research output.

More information about this series at http://www.springer.com/series/13304

Bijnan Bandyopadhyay · Abhisek K. Behera

Event-Triggered Sliding Mode Control

A New Approach to Control System Design

Springer

Bijnan Bandyopadhyay
Systems and Control Engineering
Indian Institute of Technology Bombay
Mumbai, Maharashtra
India

Abhisek K. Behera
Systems and Control Engineering
Indian Institute of Technology Bombay
Mumbai, Maharashtra
India

ISSN 2198-4182   ISSN 2198-4190   (electronic)
Studies in Systems, Decision and Control
ISBN 978-3-030-08940-5   ISBN 978-3-319-74219-9   (eBook)
https://doi.org/10.1007/978-3-319-74219-9

This Springer imprint is published by the registered company Springer International Publishing AG part
of Springer Nature
The registered company address is: Gewerbestrasse 11, 6330 Cham, Switzerland

Dedicated to my Ph.D. guide, late Dr. S. S. Lamba, former Professor, IIT Delhi
Bijnan Bandyopadhyay

Dedicated to my parents and gurus
Abhisek K. Behera

Preface

Variable structure systems using sliding mode control (SMC) was originated in USSR in the late fifties to stabilize the uncertain dynamical systems with relay as a feedback control law. It gained popularity outside USSR only after the late seventies due to a survey paper in English by Prof. Vadim I. Utkin. Since then, it has become now a well-established robust control technique to deal with the uncertainties in the plant, and achieving the system stability. A vast number of scientific publications and the practical applications of this control technique have made SMC as an important area in the control literatures.

The attention on design of SMC in discrete-time domain was paid by many researchers soon after the importance of microprocessor and computer/processors are realized in control applications in the early eighties. The first and important observation in the discrete-time design is that no exact sliding mode is achieved as in the continuous-time counterpart. A new notion of sliding mode is introduced which is known a quasi-sliding mode (QSM). This has led to the development of discrete-time SMC as an important area in SMC due to its practical importance. Many design approaches have been proposed to improve the performance of SMC for the sampled-data system.

In this monograph, a new approach to design SMC is presented using a novel implementation strategy, namely event-triggering. In this strategy, the control is updated whenever a certain stabilizing condition is violated, and hence, the system stability is always maintained. Due to the need-based control strategy, it finds a major application in spatially distributed control systems to reduce the communication among different sensor and actuator ends. So, the resources of the systems are optimally used. The event-triggering-based design of SMC not only gives the robust performance but also ensures minimal use of resources in the control system. This monograph presents the recent results on event-triggered SMC for robust stabilization of dynamical systems. In the first part of the monograph, the preliminaries on sampled-data systems with an introduction to event-triggered control are presented to familiarize the readers the event-triggering-based design of control law. In addition to this, a brief introduction to SMC and its design are also discussed. Then, the design of event-triggered SMC for both linear and nonlinear systems is given in Chaps. 2 and 3. The event-triggered SMC is presented for linear systems guaranteeing the semi-global and global stability in Chap. 2. However, only local stabilization result is discussed for the nonlinear systems, which is presented in detail in Chap. 3. In the event-triggered control, the state trajectory is continuously measured to evaluate the triggering condition which may not be practical in all applications. So, Chaps. 4 and 5 present few variants of event-triggered control, namely self-triggered and discrete event-triggered control, respectively. In the self-triggering strategy, the triggering strategy is developed without using the continuous state measurements. On the other hand, the periodic state measurements are used in discrete event-triggered control and the control is updated when it is violated at some periodic instants. In recent time, there has been a considerable amount of interest in stability of quantized control system. The final chapter presents the design of event-triggered SMC with quantized state measurements. It is our belief that the material of the monograph would serve its purpose and explore the new challenges on the topic.

This work would have not been completed without the support and encouragement from many of our friends and colleagues. The authors would like to thank Indian Institute of Technology Bombay for providing the conducive environment to carry out the research reported in the monograph. Finally, we extend our gratitude to our family for their love, support and understanding throughout the process of this endeavour.

Mumbai, India   Bijnan Bandyopadhyay
December 2017   Abhisek K. Behera

Contents

Acronyms

A-D	Analog-to-digital
D-A	Digital-to-analog
DTSM	Discrete-time sliding mode
ETCS	Event-triggered control systems
FOS	Fast output sampling
ISS	Input-to-state stable
LTI	Linear time-invariant
MIMO	Multiple-input multiple-output
MROF	Multirate output feedback
QSM	Quasi-sliding mode
SISO	Single-input single-output
SMC	Sliding mode control
VSS	Variable structure system

Symbols

$\mathbb{R}$	Set of real numbers
$\mathbb{R}_{\geq 0}$	Set of nonnegative real numbers
$\mathbb{Z}$	Set of integers
$\mathbb{Z}_{\geq 0}$	Set of nonnegative integers
$\mathbb{R}^n$	n-dimensional vector space over $\mathbb{R}$
$(a \circ b)(x)$	Composition of the two functions a and b
$x \cdot y$	Dot product of any two vectors x and y in $\mathbb{R}^n$
$P > (\geq)0$	Positive (semi)-definite matrix P
$Q^\top$	Transpose of any matrix Q
$\lvert \cdot \rvert$	Absolute value of a scalar variable '$\cdot$'
$\lVert \cdot \rVert_1$	1-norm of a finite-dimensional vector '$\cdot$'
$\lVert \cdot \rVert$	Euclidean (2-) norm of a finite-dimensional vector '$\cdot$' or a matrix '$\cdot$' of appropriate dimension
$\lambda_{\max(\min)}\{\cdot\}$	Largest (smallest) eigenvalue of a square matrix '$\cdot$'
$\nabla f(x)$	Gradient of a real-valued function $f(x)$
sup (inf)	Supremum or least upper bound (infimum or greatest lower bound)
$\lfloor x \rfloor$	Floor function that returns largest integer less than or equal to x
ln	Natural logarithm with base e ($= 2.71828$)
$\mathcal{K}$	Set of strictly increasing and continuous real-valued functions defined on the nonnegative interval with zero at zero
$\mathcal{K}_\infty$	Set of unbounded class-$\mathcal{K}$ functions
sign	Signum function
$\mathcal{F}(x)$	Set-valued map of the vector field $f(x)$ at the point of discontinuity x in Filippov's inclusion
$\overline{\mathrm{co}}$	Convex closure
$\mathcal{B}_\delta(x)$	Open ball of radius δ centred at x
$\mu(\cdot)$	Lebesgue measure of a set '$\cdot$'
ν	Sensitivity of the quantizer
M	Saturation level of the quantizer

T_i	Inter-event time/time interval between two consecutive triggering instants
τ	Constant sampling period for the discrete-time systems
N	An integer greater than or equal to the observability index of the system

Chapter 1
Introduction

This chapter briefly introduces the readers to the preliminary ideas on design and analysis of computer-controlled systems and then sliding mode control (SMC). In general, computer-controlled systems consists of both continuous and discrete-time systems that interact among themselves through the feedback channel to achieve certain objectives. Different available classical techniques, namely emulation, discrete-time and hybrid approaches, are summarized here with their own advantages and disadvantages. In almost all these techniques, the periodic sampling interval is often used to design and analyse the sampled-data systems for its simplicity and easier in design. On the other hand, aperiodic control implementation is desired in sampled-data systems to reduce the periodic computational burden and cost associated with the implementation. However, this introduces few difficulties in analysing closed loop system stability. A novel sampling strategy known as event-triggered control is introduced here where the control is updated whenever it is demanded. In this technique, the time instants for updating the control signal is determined using some rule that ensures the stability of the system. So, this strategy maintains the system stability while reducing extra burden on the system.

The design of SMC is also presented in this chapter to familiarize the readers. This is a robust controller that stabilizes the plant in the presence of external disturbances. The sliding motion and SMC are briefly elaborated to understand sliding mode with discontinuous control action. This is followed by the design of SMC for linear systems. The discrete realization of SMC, unfortunately, does not yield sliding motion exactly due to discrete nature of control is also discussed. Some control design techniques are reviewed for discrete-time sliding mode.

© Springer International Publishing AG, part of Springer Nature 2018

B. Bandyopadhyay and A. K. Behera, *Event-Triggered Sliding Mode Control*, Studies in Systems, Decision and Control 139, https://doi.org/10.1007/978-3-319-74219-9_1

1.1   Computer-Controlled Systems

Computer-controlled systems, in general, are broadly defined as a control system that allows interaction of both analog system and digital controller through computer or any other digital platform. Such systems are ubiquitous in almost all fields of engineering due to rapid advancement of digital technology. The plant represents the continuous-time system, while the control signal is a discrete in nature that is applied to the plant. The control signal is updated at the discrete instants only and, however, is held constant in between two consecutive sampling instants. This is why, it is also called as *sampled-data system*. The plant dynamics evolves in open loop manner between two discrete instants as there is no control variable. So, the interest is mainly focused on the plant behaviour at the discrete instants only leading to the so-called *discrete-time systems*. In other words, the system dynamics evolves at discrete instants with the control input applied at these instants. This is not only easy to analyse but also one of the simplest ways of implementing the control law to achieve a certain objective compared to its analog counterparts.

In actual practice, computer-controlled systems have facilitated the control design problems to a great extent by introducing the digital control signal. However, it should not be thought of as an ideal control design and implementation scenario for a physical plant. Rather, it is a limiting approach of the analog control system (both plant and control evolve in continuous-time) and all the analyses are carried out for the analog control system in sampled-data fashion. The control law is designed for any plant using system dynamics which may be defined on continuous- and/or discrete-time domain. However, in the present case only, we focus on continuous-time plant. So, the control is designed from the continuous time plant which achieves desired performance of the system when it is implemented in continuous manner. But, in discrete implementation, the continuous-time performance may also be achieved if it is implemented at a faster sampling rate. So, the computer-controlled system yet provides an alternative to analog control implementation subject to some desired performance. In spite of this, there are numerous challenges associated with computer-controlled systems that make the researchers to rethink possible new techniques for control implementation and design philosophy.

1.1.1   Basic Architecture

The basic architecture of a computer-controlled system is shown in Fig. 1.1. The plant, which evolves in continuous-time, interacts with the controller only at discrete instants. The plant states are sampled at discrete instants and converted into a digital signals. The whole process of converting analog-to-digital conversion is represented by the analog-to-digital (A-D) converter. These digital signals are then processed through a control algorithm to generate a new set of digital signals, known as discrete control signals. We see that there is same number of discrete control signals as the

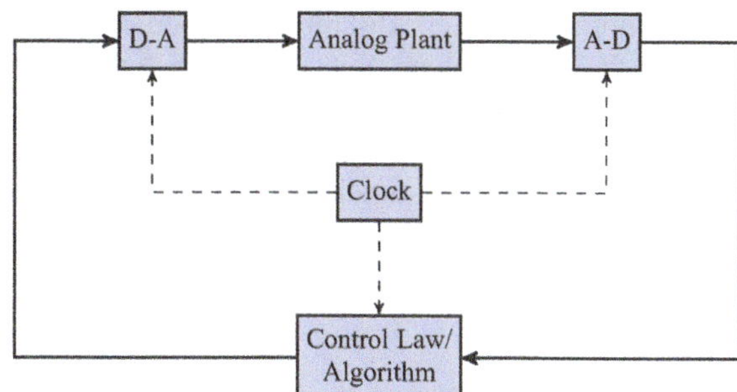

Fig. 1.1 Basic architecture of a computer-controlled systems

samples of plant states. Now, the discrete signal is processed by digital-to-analog (D-A) converter to generate a continuous-time signal between two consecutive sampling instants. But, this continuous signal is an approximation of analog control signal. The overall process from A-D to D-A is coordinated by a clock that synchronizes all the tasks carried out at A-D, control synthesis and D-A converter. Here, the control signal may be designed in the continuous-time frame work or in discrete domain depending on the design requirements.

Both the converters, often, operate in periodic time interval in the sampled-data system. The information is sampled at time instants known as *sampling instant*, and the interval between successive sampling instants is called as *sampling period*. For every sampling instant, the control signal is computed and the same is applied to the plant by its approximate analog signal through D-A converter. The control signal is held constant in every sampling interval making the system as open loop, and hence, the system evolves in open loop manner. This is one of the differences that makes the computer-controlled systems different from other feedback systems. There has been many techniques available to explore the analysis and design of such system. But, yet the full potential of this system needs to be investigated for effective use of computer-controlled systems. For instance, the effect of sampling interval on system performance, the absence of clock that synchronizes both A-D and D-A converters, etc. are need to be investigated. Of all these, sampling interval variation is one of the most important and challenging problems in computer-controlled system. Too fast sampling is sometime unnecessary, while the slow sampling may deteriorate the system performance. So, it is always desirable to have an optimal sampling interval that stabilizes the computer-controlled system.

1.1.2 Design Techniques

Many techniques have been used to analyse and design of computer-controlled systems. The closed loop system is hybrid in nature involving both continuous and discrete dynamics; however, the standard available technique deals with either continuous-time design or discrete-time design. That means design the control either using continuous-time or discrete-time model and apply it to the continuous-time plant. In the same manner, the stability analysis of the system is carried out using either of these models depending on control signal. Broadly speaking, there are three methods available for analysing the computer-controller system, namely emulation-based approach, discrete-time approach and hybrid system approach. Also, in most applications the constant sampling period is used for implementing the discrete controller.

1.1.2.1 Emulation-Based Approach

It may also be seen as continuous-time approach design to sampled-data system. Here, the performance specifications are in continuous-time domain since the plant and controller both are in continuous-time domain. The basic idea is first to ignore the A-D and D-A converters in Fig. 1.1 and follow the design steps of any stabilizing controller. The continuous-time controller is then approximated by replacing derivatives with finite-differences and continuous-time signals with sampled values at that instants. Though it is an approximation, the satisfactory performance of the sampled-data system is still achieved by this approximation. Here, the controller designed from continuous-time plant *emulates* the behaviour of continuous-time plant in spite of discrete implementations, and hence, this is known as *emulation-based approach*. However, the main issue in this approach is that the sampling period is not taken into account in the design of controller. It is very natural that the emulated controller gives the stability of sampled-data system for a range of sampling period only while it destabilizes for other values of sampling period.

1.1.2.2 Discrete-Time Approach

This approach is simple and is based on the discrete-time model of the plant. The continuous plant interacts with digital controller through A-D and D-A converters. Thus, the controller sees the plant as discrete-time model through these converters. The discrete-time representation of the plant is obtained by combining the plant with A-D and D-A converters. There are numerous approaches available for discretizing the plant for a given sampling period, and the popular among them are Euler discretization, zero-order hold (ZOH) discretization, etc. The closed loop response is analysed only at discrete instants since it ignores the inter-sampling behaviour of the plant. On the other hand, the stability of the system is analysed by ignoring the dynamics between two sampling instants. The wide application of this technique is found in many slow processes where it is enough to study the system behavior only at some periodic intervals. However, the well-known shortcomings of this technique are that no inter-sampling dynamics of the plant can be analysed and selection of a suitable sampling interval that captures the undesirable phenomenon in the system.

1.1.2.3 Hybrid Approach

As the name suggests, in this technique, both the continuous and discrete behaviours are analysed for the sampled-data system without representing the system by some approximated system dynamics. This is why, it is also known as direct design approach to sampled-data system. Due to this hybrid nature of the system, the design and stability methods are complicated than earlier two methods.

It is to be noted that variable sampling period may also be used for designing the controller for sampled-data systems. However, it involves many design issues for analysing the stability due to restricted mathematical tools and/or time-varying nature of system dynamics. Nevertheless, many attempts have been made for stabilizing the sampled-data systems with aperiodic control sampling process. Event-triggering strategy is one of such techniques that generates nonuniform sampling instants while ensuring system stability. In this book, only event-triggered technique will be emphasized for computer-controlled systems.

1.2 Event-Triggered Control

Event-triggering strategy is a control implementation technique that is motivated from the Lebesgue sampling. In this case, sampling period is not constant but is determined by the evolution of system trajectory satisfying some stability condition. To understand the concept of event-triggered control, first Lebesgue sampling technique is discussed.

In classical sampled-data system, generally constant sampling period is chosen and the control is implemented once this constant time period is elapsed. This is generally referred as Reimann sampling as shown in Fig. 1.2a. For any constant $h_R >$ 0, the sampling of the continuous-time signal $\psi(t)$ takes place at every h_R intervals of time. It does not monitor the evolution of signal $\psi(t)$. As a result of this, the important concern in Reimann sampling is the proper selection of sampling period to capture the transient behaviour of the plant. On the other hand, the signal may also be sampled at the time instants when state evolution from its immediate past sampled value crosses a certain threshold value (say, h_L) as the time does in Reimann sampling. This is known as Lebesgue sampling which is shown in Fig. 1.2b. Though both the sampling techniques are different, these have similarity in the sampling mechanism. In the former, the time is measured while state is monitored in the latter case. However, in doing so many advantages are obtained in the case of Lebesgue sampling. For example, it is not necessary most of the time to update the control signal frequently at periodic interval. So, Lebesgue sampling gives sampling instant whenever it is needed subject to some satisfactory system performance.

In case the function $\psi(t)$ is a finite-dimensional vector for any fixed t, the sampling instant may be decided by observing the individual state evolution of the vector-valued function. Then, for any given constants $h_{L_i} > 0$ with $i = 1, 2, \ldots, n$, the individual state evolution, denoted by $h_{L_i}(t)$, is observed for generating sampling

Fig. 1.2 Comparison of
Reimann and Lebesgue
samplings

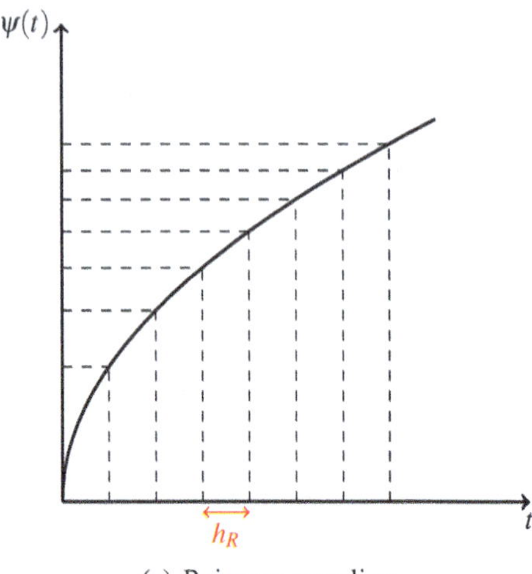

(a) Reimann sampling

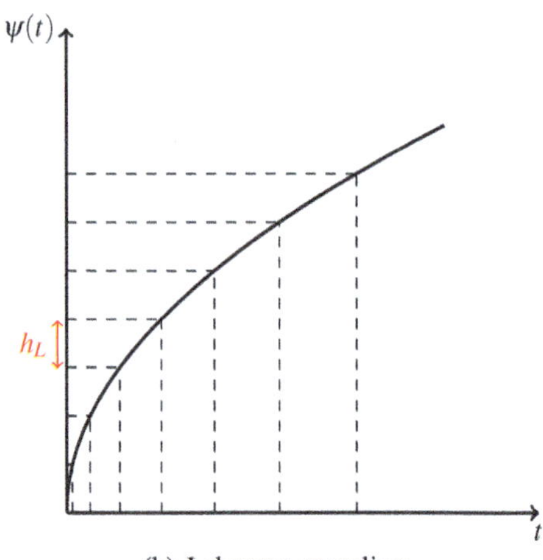

(b) Lebesgue sampling

instant of the corresponding state. However, it is very complicated and difficult to
analyse the stability of closed loop system. Another school of thought is to sample all
the states simultaneously whenever certain condition is violated. Thus, this strategy
does not necessitate individual sampling of the state at different time instants. Due to
this, the latter is more convenient for implementing practically than the earlier one.

1.2.1 Preliminary Idea

Event-triggered control is one of such techniques that generates the sampling instant (also called as triggering instant) for sampling and updating the control signal. To provide a preliminary idea on event-triggered control, we consider a nonlinear dynamical system

$$\dot{x} = f(x, u), \qquad x(0) = x_0 \in \mathbb{R}^n \tag{1.1}$$

where the function $f(\cdot, \cdot)$ is Lipschitz with respect to both the arguments $u \in \mathbb{R}^m$. Let there exists a continuous feedback control law $u(x) = \pi(x)$ such that the dynamics

$$\dot{x} = f(x, \pi(x))$$

is asymptotically stable. It is assumed that the control is implemented digitally to the plant. So, the control signal $\pi(x)$ is computed for every sampling instant and is applied to the plant at these discrete instants. Then, the system becomes open loop between two consecutive sampling instants. However, due to this, the discrete error is introduced in the plant, defined by $e(t) = x(t_i) - x(t)$ with $e(t_i) = x(t_i) - x(t_i) = 0$ where $t \in [t_i, t_{i+1})$. This error appears in the plant due to discrete implementation of continuous-time control, but its value is zero if the control is continuously updated as in analog implementation.

Further, we assume that the system (1.1) is input-to-state stable (ISS) with respect to the error $e(t)$. That means there exists a continuously differentiable Lyapunov function $V : \mathbb{R}^n \to \mathbb{R}_{\geq 0}$ such that

$$\underline{a}(\|x\|) \leq V(x) \leq \overline{a}(\|x\|) \tag{1.2}$$

$$\nabla V(x) \cdot f(x, \pi(x + e)) < -a(\|x\|) + \gamma(\|e\|) \tag{1.3}$$

for some class-$\mathcal{K}_\infty$[1] functions $\underline{a}$, $\overline{a}$, a, and class-$\mathcal{K}$ function γ. Here, the notation '·' denotes inner (scalar) product. Event-triggering strategy is developed for determining the sampling instants such that desired stability is achieved. In this case, the asymptotic stability of the system is desired with the discrete implementation of the control law. So, the obvious condition for which this holds is $\gamma(\|e\|) < \sigma a(\|x\|)$ for some $\sigma \in (0, 1)$. This can be simplified further, by assuming a^{-1} and γ are Lipschitz on some compacts, as $\sigma\|x\| > L_e\|e\|$, where L_e is an appropriate constant. Thus, the triggering instant may be generated by executing the following,

$$t_{i+1} = \inf\left\{t > t_i : \sigma\|x(t)\| \leq L_e\|e(t)\|\right\}. \tag{1.4}$$

---

[1] Any function a is said to be class-$\mathcal{K}$ if it is continuous, strictly increasing, zero at zero. Again, it is said to be class-$\mathcal{K}_\infty$ if it belongs to class-$\mathcal{K}$ and is unbounded. Clearly, class-$\mathcal{K}_\infty$ functions are subsets of class-$\mathcal{K}$ functions.

This is known as triggering rule for event-triggered control $\pi(x)$. It is seen that this ensures $L_e \|e\| < \sigma \|x\|$ which implies that $\gamma(\|e\|) < \sigma a(\|x\|)$ also holds. This implies from (1.2) and (1.3) that

$$
\begin{aligned}
\dot{V} &< -(1-\sigma)a(\|x\|) \\
&\leq -(1-\sigma)\left(a \circ \overline{a}^{-1}\right)(V) \\
&= -(1-\sigma)a^{\star}(V) \\
&< 0
\end{aligned}
$$

where $\mathscr{K}_{\infty} \ni a^{\star} := a \circ \overline{a}^{-1}$ which is the composition of two functions a and $\overline{a}^{-1}$. This shows that closed loop system is asymptotically stable with the control applied at discrete instants. Moreover, in the event-triggering technique the inter-sampling behaviour is considered for stability of the closed loop system.

It is worth mentioning here that the event-triggered control scheme not only guarantees system stability but also ensures the convergence of inter-sampling behaviour. From the above, it is observed that the Lyapunov function decreases continuously and goes to zero as the time tends to infinity. This is one of the important properties of the event-triggered control system (ETCS).

1.2.2 Stability of Event-Triggered Systems

In this technique, the triggering instants are generated by the triggering rule at which the control signal is updated and this results the closed loop system stability. However, it might happen that the control signal is not updated at the triggering instant when triggering instants are too close to each other. This demands fast execution of control tasks, or even in worst-case continuous-time like execution which is not possible by digital processor. It may be noted here that in periodic execution such situation does not arise due to each sampling/triggering instant that occurs after every constant sampling period.

Let $\{t_i\}_{i \in \mathbb{Z}_{\geq 0}}$ be sequence of triggering instants generated by some stabilizing triggering rule. We define $T_i = t_{i+1} - t_i$ as the *inter-event/execution time* for any given triggering sequence $\{t_i\}_{i \in \mathbb{Z}_{\geq 0}}$. For stability of the event-triggered system, the inter-event time must be strictly greater than zero, i.e. $T_i > a_T$ for all $i \in \mathbb{Z}_{\geq 0}$ and some positive constant a_T. This guarantees the *Zeno*-free execution of triggering sequence. The positive inter-event time ensures control is updated after every finite-time interval only. This is essential for the processor to execute the control task and update the control signal. In other words, it can be said that $\{t_i\}_{i \in \mathbb{Z}_{\geq 0}}$ is an increasing sequence, i.e. $t_0 < t_1 < t_2 < \cdots$ such that $t_{i+1} > t_i + a_T$. Such a triggering sequence is feasible for implementing the control practically to ensure the stability of closed loop system. The triggering instants generated by some triggering rule that is not necessarily satisfying the above property would make the event-triggered system unstable.

Example 1.1 Consider a scalar nonlinear control system as

$$\dot{x} = x^2 + u$$

where $x \in \{[-c, c] : c \in \mathbb{R}_{>0}\}$ which is a compact set. Any stabilizing controller can be designed for the above system to ensure the asymptotic stability. Let $u = -x^2 - kx$ be a feedback control which ensures the asymptotic stability of the system with $k > 0$. This control is applied to the above system at discrete instants only such that closed loop system is stable. So, the discrete-time control is given as

$$u(t) = -x^2(t_i) - kx(t_i), \qquad t \in [t_i, t_{i+1}), \quad i \in \mathbb{Z}_{\geq 0}.$$

It can be shown that the closed loop system with the above discrete control is ISS with respect to the error. Choose $V(x) = \frac{1}{2}x^2$. Then, with some calculation, we arrive at

$$\begin{aligned}
\dot{V}(x(t)) &= x(t)\left(x^2(t) - x^2(t_i) - kx(t_i)\right) \\
&= -x(t)\left(x(t) + x(t_i)\right)e(t) - kx(t)x(t_i) \\
&\leq 2c|e(t)||x(t)| - kx(t)(e(t) + x(t)) \\
&= 2c|e(t)||x(t)| - kx(t)e(t) - kx^2(t) \\
&\leq 2c|e(t)||x(t)| + k|x(t)||e(t)| - kx^2(t) \\
&\leq (2c + k)|e(t)||x(t)| - kx^2(t).
\end{aligned}$$

Here, we use the fact $|x| \leq c$ and $x(t_i) = e(t) + x(t)$. Now applying Young's inequality[2] to the first term ($\varepsilon = \frac{2c+k}{k}$), we obtain

$$\begin{aligned}
\dot{V}(t) &\leq -\frac{k}{2}|x(t)|^2 + \frac{(2c + k)^2}{2k}|e(t)|^2 \\
&= -a(|x(t)|) + \gamma(|e(t)|)
\end{aligned}$$

where $a(r)$ and $\gamma(r)$ are given as

$$a(r) = \frac{k}{2}r^2 \quad \text{and} \quad \gamma(r) = \frac{(2c + k)^2}{2k}r^2.$$

Thus, the triggering rule is designed according to (1.4) which stabilizes the system and is given by

---

[2] Young's inequality for exponent two states that for any nonnegative real numbers p, q and every $\varepsilon > 0$, the following holds

$$pq \leq \frac{p^2}{2\varepsilon} + \frac{\varepsilon q^2}{2}.$$

$$t_{i+1} = \inf\left\{t > t_i : \sigma \frac{k}{2c+k}|x(t)| \le |e(t)|\right\}$$

for some $\sigma \in (0, 1)$. This triggering rule ensures $|e(t)| < \sigma\frac{k}{2c+k}|x(t)|$ for all time and thus implies that $\dot{V} < 0$ for all time. Hence, the closed loop system becomes asymptotically stable even if the control is applied at the discrete instants generated by the triggering sequence $\{t_i\}_{i\in\mathbb{Z}_{\ge 0}}$. It can also be established that the triggering rule does not have a Zeno triggering sequence. For $x \in \{[-c, c] : c \in \mathbb{R}_{>0}\}\backslash\{0\}$, one can write

$$\begin{aligned}
\frac{\mathrm{d}}{\mathrm{d}t}\frac{|e(t)|}{|x(t)|} &= \frac{1}{x^2(t)}\left(\frac{\mathrm{d}}{\mathrm{d}t}|e(t)|\,|x(t)| - \frac{\mathrm{d}}{\mathrm{d}t}|x(t)|\,|e(t)|\right) \\
&\le \frac{1}{x^2(t)}\left(\left|\frac{\mathrm{d}}{\mathrm{d}t}e(t)\right||x(t)| - \frac{\mathrm{d}}{\mathrm{d}t}|x(t)|\,|e(t)|\right) \\
&\le \frac{1}{x^2(t)}\left(\left|\frac{\mathrm{d}}{\mathrm{d}t}x(t)\right||x(t)| + \left|\frac{\mathrm{d}}{\mathrm{d}t}x(t)\right||e(t)|\right) \\
&= \frac{1}{x^2(t)}(|x(t)| + |e(t)|)\left|\frac{\mathrm{d}}{\mathrm{d}t}x(t)\right|.
\end{aligned}$$

Now, using the fact $|x| \le c$ and the control law $u(t) = -x^2(t_i) - kx(t_i)$ for all $t \in [t_i, t_{i+1})$ and $i \in \mathbb{Z}_{\ge 0}$, one can easily obtain that

$$\begin{aligned}
\left|\frac{\mathrm{d}}{\mathrm{d}t}x(t)\right| &= \left|x^2(t) - x^2(t_i) - kx(t_i)\right| \\
&= |-(x(t) + x(t_i))\,e(t) - k\,(e(t) + x(t))| \\
&= |x(t) + x(t_i)|\,|e(t)| + k\,|e(t) + x(t)| \\
&\le 2c|e(t)| + k|e(t)| + k|x(t)| \\
&< (2c + k)|e(t)| + (2c + k)|x(t)| \\
&= (2c + k)(|e(t)| + |x(t)|).
\end{aligned}$$

Using this in the following, it can be written as

$$\begin{aligned}
\frac{\mathrm{d}}{\mathrm{d}t}\frac{|e(t)|}{|x(t)|} &< \frac{1}{x^2(t)}(|x(t)| + |e(t)|)^2\,(2c + k) \\
&= (2c + k)\left(\frac{|e(t)|}{|x(t)|} + 1\right)^2.
\end{aligned}$$

The solution to the above differential inequality can be obtained using comparison Lemma. Then, the solution to the above differential can be obtained as

$$\frac{|e(t)|}{|x(t)|} \le \mu(t), \qquad t \in [t_i, t_{i+1})$$

where $\mu(t)$ satisfies the differential equation $\dot{\mu} = (2c + k)(1 + \mu)^2$ with the initial condition $\frac{|e(t_i)|}{|x(t_i)|} = \mu(t_i) = 0$. Then, corresponding to triggering mechanism for this system, the lower bound of the inter-event time is obtained as

$$T_i > \frac{\sigma k}{(2c + (1 + \sigma)k)(2c + k)} > 0.$$

This shows that the inter-event time is lower bounded by a positive quantity which is strictly greater than zero. Indeed, it is necessary to eliminate the Zeno execution of triggering sequence and ultimately to guarantee the system stability.

In the numerical simulation, the values of c, k and σ are selected as 5, 1 and 0.85, respectively. The initial condition is taken as $x(0) = 4$. The response of the system is shown in Fig. 1.3. It is seen that state trajectory goes to zero as time tends to infinity. The varying sampling interval or inter-event time generated by executing the triggering rule is shown in Fig. 1.4. It is seen that the inter-event time is lower bounded from zero which is given by 0.0065 and it increases to a value as high as 0.072. The plot of control signal is also shown in Fig. 1.5. As the sampling interval increases, the control signal also remains constant until the next sampling instant is generated. The plot of Lyapunov function is given in Fig. 1.6. It is seen that event-triggered control implementation achieves asymptotic stability of the closed loop system with guaranteed convergence of inter-sampling behaviour of the system.   □

1.2.3 Need of Event-Triggered Control

Event-triggered control is one of the aperiodic control implementation strategies in digital platform that ensures closed loop system stability. Unlike periodic sampling,

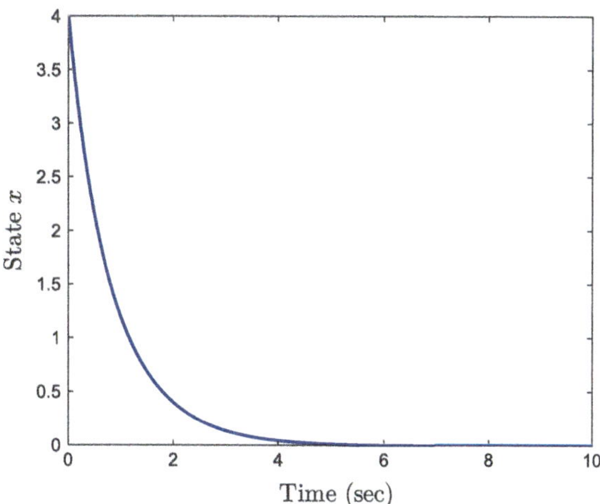

Fig. 1.3 Response of the system

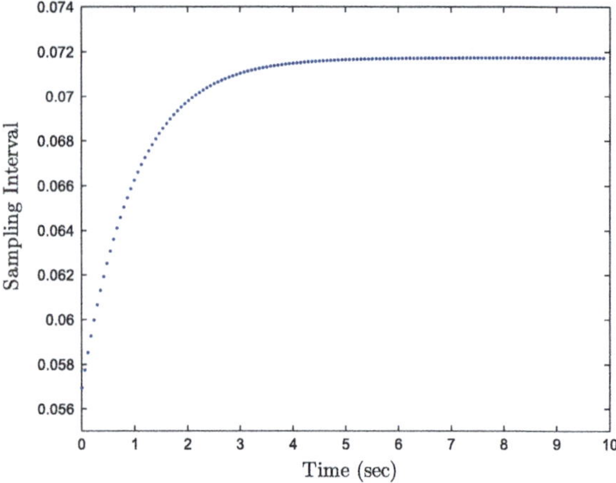

Fig. 1.4 Variation of sampling interval or inter-event time

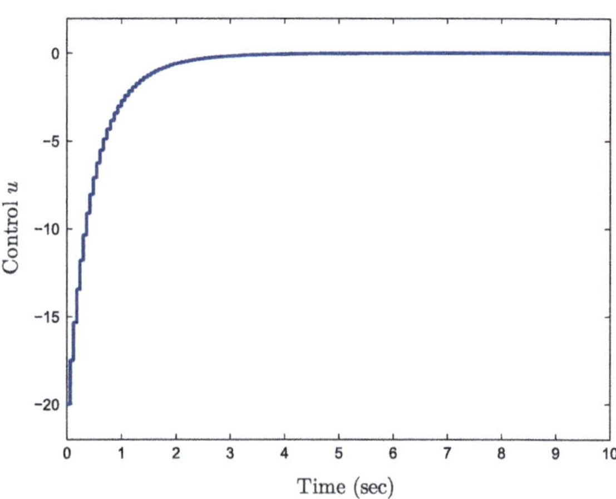

Fig. 1.5 Event-triggered control signal

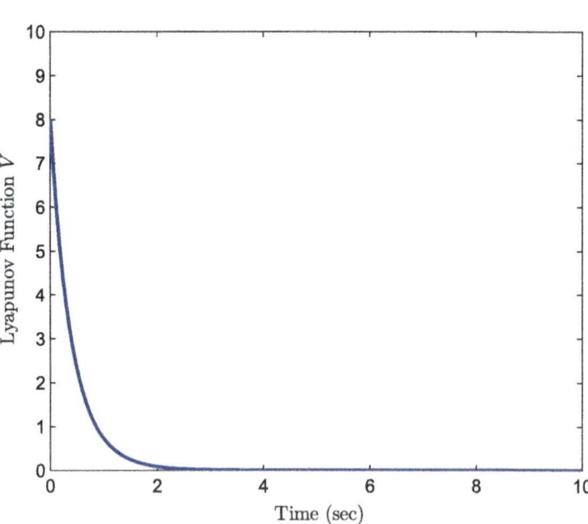

Fig. 1.6 Time evolution of Lyapunov function

here sampling instants are determined whenever it is demanded subject to system stability. Otherwise, no control signal is updated. So, if there is no triggering for long duration, then the control is not updated while maintaining the stability of the system. Such a need-based strategy is more useful in resource-constrained systems such as networked control system, embedded control systems. In these systems, frequent state transmission in periodic interval is sometimes not desirable due to bandwidth constraints of the communication channel.

In the event-triggered control strategy, there is less computational burden than that of periodic control implementation subject to some satisfactory system performance. This is mainly because the control signal is computed only when an event is triggered. This is more appealing in many practical applications to avoid control computation whenever the states do not change rapidly. Moreover, it enables the digital processors to work in parallel scheduling tasks by minimizing the periodic execution.

Another important aspect of event-based technique is the scheduling of sensors in practical systems. Sensors are the integral part of the control systems which measure the state/output trajectory and transmit it to the control end. In case the data is to be transmitted over lossy channel, it is desirable to have a control policy that requires minimal state transmission. So, in such cases, scheduling of sensors is so designed that it can decide the transmission instant. Event-triggering strategy is very much helpful in scheduling these sensors.

Apart from these, this strategy is successfully deployed in many other applications such as in signal processing, state estimation. However, the discussion of these topics is the beyond the scope of this book, so here only the control aspect of event-triggered technique is exploited.

1.3 Sliding Mode: An Introduction

Sliding mode has its root in the variable structure system (VSS) where the system structure changes during the evolution of the system dynamics. In VSS, the structure of the system is changed or switched such that the closed loop system is asymptotically stable. However, in SMC the system trajectory is forced to remain on a predesigned manifold by the action of discontinuous control signal. In this case, the vector fields on both the sides of sliding (we also refer it as 'switching') manifold act towards this manifold and thus maintain the trajectory along it. This eventually ensures sliding takes place and is called *sliding mode*. Since the system dynamics switches between two structures while *sliding mode* is enforced, SMC is referred as a special class of VSS where switching takes place along a predesigned switching manifold. Due to this, overall system becomes discontinuous on this manifold. It is not necessarily required that sliding mode is enforced in the system from the very beginning, but it must start in some finite-time. Otherwise, system cannot be said to be in *sliding mode*. In controlled dynamical system, the control law is often designed that brings *sliding mode* in the system and is called as *sliding mode control*. This control is responsible for sliding mode.

To illustrate the concept of sliding mode, consider a controlled nonlinear dynamical system as

$$\dot{x} = f(x, u), \qquad x \in \mathbb{R}^n \quad \text{and} \quad u \in \mathbb{R}. \tag{1.5}$$

The vector field $f : \mathbb{R}^n \times \mathbb{R} \to \mathbb{R}^n$ given in (1.5) is Lipschitz with respect to its first argument. Let the sliding manifold be given for any continuous switching function, denoted by $s(x)$ which maps $\mathbb{R}^n$ to $\mathbb{R}$, as

$$\mathscr{S} = \left\{ x \in \mathbb{R}^n : s(x) = 0 \right\}.$$

Similarly, we also define $\mathscr{S}^+ = \{x \in \mathbb{R}^n : s(x) > 0\}$ and $\mathscr{S}^- = \{x \in \mathbb{R}^n : s(x) < 0\}$. The control law that brings sliding mode in the system (1.5) is given by

$$u(x) = \begin{cases} u^+(x) & \text{if } s(x) > 0, \\ u^-(x) & \text{if } s(x) < 0. \end{cases} \tag{1.6}$$

The control functions u^+ and u^- are continuous, and also $u^+ \neq u^-$. Clearly, this implies that the control u is discontinuous on $s = 0$. The corresponding resulting vector fields are $f^+(x, u^+)$ and $f^-(x, u^-)$ due to the control signals u^+ and u^-, respectively. So, the vector field $f(\cdot, \cdot)$ is also discontinuous on $\mathscr{S}$. The existence of solutions of the closed loop system (1.5) with control law (1.6) cannot be explained using the classical existence theorem. Indeed, in this case, the closed loop system becomes discontinuous and is referred as system with discontinuous right-hand side. Though there are many techniques available for defining solutions of such systems, in this book we understood the solutions of the system in Filippov's sense [1].

1.3.1 Dynamics During Sliding Mode

The existence of solutions of the dynamical system (1.5) is briefly discussed in Filippov's sense. The dynamical system is first replaced by a set-valued function, called differential inclusion, on the zero (Lebesgue) measure set where solution is not defined in classical sense. We denote it by $\mathscr{F}(x, u)$. Thus, the differential inclusion for the system (1.5) is written as

$$\dot{x} \in \mathscr{F}(x, u). \tag{1.7}$$

The set-valued map $\mathscr{F}$ coincides with f; i.e., $\mathscr{F}$ contains only one element f, whenever the latter is continuous on its respective domain. In other words, the set-valued map represents the vector fields of the dynamical system at the points of discontinuity. One of the main reasons to replace the dynamical equation (1.5) by an inclusion (1.7) is that to capture all the vector fields of the system at the point of discontinuity. Another way of interpretation is that the right-hand side is enlarged such that all vector fields in the vicinity of manifold are contained in the inclusion. Once that is accomplished, the most obvious question is under what conditions the solution of the system (1.7) exists. Before that, we briefly discuss how the set-valued map can be constructed from an dynamical system (1.5) that has discontinuous right-hand side.

Since the inclusion, given in (1.7), captures all the vector fields in some sufficiently small δ-neighbourhood of $s(x) = 0$, the set-valued map is constructed by collecting the convex combination of the vector fields in that neighbourhood. Thus, the limiting vector fields in the small neighbourhood of the domains $\mathscr{S}^+$ and $\mathscr{S}^-$ for any $x \in \mathscr{S}$ are obtained as

$$\lim_{\substack{y \in \mathscr{S}^+ \\ y \to x}} f(y, u^+(y)) = f^+(x, u^+(x)), \qquad \lim_{\substack{y \in \mathscr{S}^- \\ y \to x}} f(y, u^-(y)) = f^-(x, u^-(x)).$$

Then, the set-valued map $\mathscr{F}(x, u)$ is obtained by collecting all the vector fields pointing the line segment joining the end points of the vector fields in $\mathscr{S}^+$ and $\mathscr{S}^-$, i.e. f^+ and f^-. Note that every trajectory in $\mathscr{S}^+$ crosses $\mathscr{S}$ before reaching $\mathscr{S}^-$ and vice versa. This implies that the line segment joining f^+ and f^- also intersects $\mathscr{S}$. If $T_x\mathscr{S}$ represents the tangent vector of $\mathscr{S}$ at the point x, then the line segment also intersects $T_x\mathscr{S}$ at some point. Thus, this intersection point gives the end point of the vector $f^0(x, u)$ such that

$$\dot{x} = f^0(x, u) \tag{1.8}$$

holds for the point $x \in \mathscr{S}$. Similarly, it can be defined for other points on $\mathscr{S}$. This gives the motion of the system trajectory on the sliding manifold. In other words, the function $x(t)$ satisfies (1.8) is a solution of the system (1.5). This is because this solution also satisfies the inclusion (1.7) as f^0 is contained in the inclusion. Also, $f^- \neq f^0$ and $f^+ \neq f^0$; the motion of the system remains tangent to $\mathscr{S}$ and is called *sliding motion*. Thus, during sliding mode, the system trajectory is a solution to (1.8) which is also a solution of (1.5) due to (1.7).

Now, we discuss how to find the velocity vector f^0 during sliding mode. Define f_N^+ and f_N^- as the projections of the vectors f^+ and f^-, respectively, onto the normal to the surface $\mathscr{S}$ at the point $x \in \mathscr{S}$. Then, the vector field f^0 is calculated using (1.8) as

$$f^0 = \theta f^+ + (1 - \theta)f^-, \qquad \theta = \frac{f_N^-}{f_N^- - f_N^+}, \quad \theta \in [0, 1]. \tag{1.9}$$

This shows the vector field during sliding mode is a convex combination of the vector fields f^+ and f^- such that the trajectory is tangent to surface $\mathscr{S}$. The value of θ given in (1.9) is obtained from the relation

$$f_N^0 = \theta f_N^+ + (1 - \theta)f_N^- = 0.$$

That means the velocity vector f^0 is along $T_x\mathscr{S}$. The normal components f_N^- and f_N^+ can also be represented in terms of the switching function $s(x)$. It is to be noted that $\nabla s(x)$ represents normal to the surface $\mathscr{S}$ at the point $x \in \mathscr{S}$. Then,

$$f_N^+ = \frac{\nabla s \cdot f^+}{\|\nabla s\|}, \qquad f_N^- = \frac{\nabla s \cdot f^-}{\|\nabla s\|}, \quad \text{and} \quad \theta = \frac{\nabla s \cdot f^-}{\nabla s \cdot (f^- - f^+)}$$

where the notation '$\cdot$' denotes inner (scalar) product; i.e., for any two vectors z_1 and z_2 in $\mathbb{R}^n$, $z_1 \cdot z_2 = \langle z_1, z_2 \rangle = z_1^\top z_2$. Thus, Filippov's differential inclusion characterizes sliding motion of the dynamical system (1.5) by constructing the set-valued map at the point of discontinuity, and the solution is obtained such that vector filed is tangential to the sliding manifold $\mathscr{S}$.

Apart from these, there is an another definition due to Utkin that also explains system dynamics during sliding motion. According to Utkin, during sliding motion there exists an *equivalent control*, u^{eq}, such that the vector $f(x, u^{\text{eq}})$ is tangent to $\mathscr{S}$ at $x \in \mathscr{S}$. In other words, one has

$$\nabla s \cdot f(x, u^{\text{eq}}) = 0.$$

The solution to the system is an absolutely continuous function that satisfies the above relation for almost all time. Any function $x(t)$, such that $\dot{x} = f(x, u^{\text{eq}})$ and above relation holds for almost all time, is also a solution to (1.5). Since the vectors on either sides of $\mathscr{S}$ are pointing towards $\mathscr{S}$, the trajectory/solution never leaves the surface $\mathscr{S}$. This is popularly known as *equivalent control method*. The vector field $f(x, u^{\text{eq}})$ can be computed as follows. For a fixed $x \in \mathscr{S}$, plot the locus $f(x, u)$ by varying the control u from $u^-(x)$ to $u^+(x)$. Then, the vector field corresponding to the equivalent control is also on this arc. This vector field $f(x, u^{\text{eq}})$ is tangential to sliding manifold $\mathscr{S}$ at x and can be drawn from the point $x \in \mathscr{S}$ to the point on the locus. Here, the locus may be considered as a set-valued map in the equivalent control method.

One of major differences between Filippov's inclusion and Utkin's equivalent control method is that in the latter the set-valued map is obtained by varying u from $u^-(x)$ to $u^+(x)$ for a fixed $x \in \mathscr{S}$. However, both these techniques coincide when the end point of the vector $f(x, u^{\text{eq}})$ remains on the line segment joining $f^-(x, u^-)$ and $f^+(x, u^+)$. Again apart from this case, also for the systems linear with respect to control, the set-valued map is a straight line joining $f^-(x, u^-)$ and $f^+(x, u^+)$ in both the equivalent control method and Filippov's method and thus yields same result for vector fields on the discontinuous manifold, i.e. $f(x, u^{\text{eq}}) = f^0(x, u)$.

So, throughout the book, we understood the sliding motion in Filippov's sense. Here, the set-valued map $\mathscr{F}(x, u)$ is more formally defined. It is the smallest convex set containing all the limiting values of vector function $f(y, u(y))$ when y in δ-neighbourhood of x in $\mathscr{S}$ tends to x such that $y \notin \mathscr{S}$. This is given as

$$\mathscr{F}(x, u(x)) = \bigcap_{\delta > 0} \bigcap_{\mu N = 0} \overline{\mathrm{co}} f\left(\mathscr{B}_\delta(x) \backslash N, u(\mathscr{B}_\delta(x) \backslash N)\right) \qquad (1.10)$$

where $\overline{\mathrm{co}}$ is the convex closure, μN is the Lebesgue measure of set N, and $\mathscr{B}_\delta(x)$ denotes a open ball of radius $\delta > 0$ centred at $x \in \mathscr{S}$. Some conditions are stated by the Filippov under which the solution for sliding motion exists and these are called basic conditions.

Definition 1.1 ([1], 2, *Sect. 7, p. 76*) The set-valued map $\mathscr{F}(x, u(x))$ is said to satisfy the *basic conditions* if it is *nonempty, bounded and closed, convex*, and the function $\mathscr{F}(x, u(x))$ is *upper semicontinuous* in x and u.

Then, if the basic condition holds for differential inclusion, there exists an absolutely continuous function $\xi(t)$ for $t \in I$ that satisfies Filippov's inclusion

$$\dot{\xi}(t) \in \mathscr{F}(\xi(t), u(\xi(t)))$$

for almost everywhere on I. This is called the solution to (1.7) and hence to (1.5). Throughout the book, whenever the solution is not defined in usual sense, we understood its Filippov's sense, i.e. an absolutely continuous signal that satisfies the Filippov's inclusion.

1.3.2 Design of Sliding Mode Control

Here, the design of SMC is discussed that brings the sliding mode in the system. We consider a linear time-invariant (LTI) system to explain the sliding mode for simplicity, and also, the most part of this book deals with same. Consider a LTI system as

$$\dot{x} = Ax + B(u + d) \qquad (1.11)$$

where state $x \in \mathbb{R}^n$ and scalar control input $u \in \mathbb{R}$. The matrices A and B are of appropriate dimensions. Here, d is the external disturbance that enters the system through the input channel which is called as *matched* uncertainty. The following assumption on the disturbance holds throughout the book.

Assumption 1.1 The disturbance $d(t)$ is bounded with known upper bound. That means there exists d_0 such that $\sup_{t \geq 0} |d(t)| \leq d_0$.

The major interest of SMC is that it yields sliding motion in the presence of disturbance d. First, write the system (1.11) in regular form as given below

$$\dot{x}_1 = A_{11}x_1 + A_{12}x_2 \tag{1.12}$$

$$\dot{x}_2 = A_{21}x_1 + A_{22}x_2 + B_2(u+d) \tag{1.13}$$

where $x_1 \in \mathbb{R}^{n-1}$ and $x_2 \in \mathbb{R}$. If the system (1.11) is not in regular form, then a nonsingular transformation T can be defined such that it transforms the system (1.11) into (1.12)–(1.13) with the transformation $z = Tx$. So, without loss of any generality, we assume that the original system is in regular form. Design the switching function or sliding variable as $s(x) = c^{\top}x$ where $c = [c_1^{\top} \ 1]^{\top}$. Then, the sliding manifold is rewritten as

$$\mathscr{S} = \left\{ x \in \mathbb{R}^n : s(x) = c^{\top}x = 0 \right\}. \tag{1.14}$$

The surface parameter c_1 is selected such that $A_{11} - A_{12}c_1^{\top}$ is Hurwitz. This can be easily done for LTI system using any pole placement techniques, provided the pair (A_{11}, A_{12}) is controllable. It is seen that (A, B) is a controllable pair if and only if (A_{11}, A_{12}) is controllable. The objective is to ensure the sliding mode in the system (1.11) in finite-time. We will now design the control law to bring the sliding motion in the system. Differentiating $s(x)$ with respect to time, we write

$$\dot{s} = c^{\top}Ax + c^{\top}Bu + c^{\top}Bd.$$

The control law which brings sliding mode is given as

$$u = -(c^{\top}B)^{-1}\left(c^{\top}Ax + K\mathrm{sign}s\right) \tag{1.15}$$

where $K > \sup_{t \geq 0} |c^{\top}Bd(t)|$. The vector c is designed such that $c^{\top}B$ is invertible. Here, sign is the signum function and is defined for any variable $\varphi \in \mathbb{R}$ as

$$\mathrm{sign}\varphi := \begin{cases} 1 & \text{if } \varphi > 0, \\ [-1, 1] & \text{if } \varphi = 0, \\ -1 & \text{if } \varphi < 0. \end{cases}$$

This definition of sign allows one to construct the set-valued map at the point of discontinuity. Now, substituting the SMC (1.15) in the sliding dynamics yields

$$\dot{s} = -K\mathrm{sign}s + c^{\top}Bd. \tag{1.16}$$

In order to ensure the system trajectory reaches the manifold $\mathscr{S}$ in finite-time, following must hold

$$s\dot{s} \leq -\eta|s|$$

for some $\eta > 0$ which is popularly known as η-*reachability* condition in the sliding mode literature. This condition ensures reachability to the sliding surface in finite-time. Also, the finite-time reachability to sliding surface can also be seen by Lyapunov analysis. Choose Lyapunov function $V = \frac{1}{2}s^2$. Then, differentiating V along the solutions of the system (1.11), one obtains

$$
\begin{aligned}
\dot{V} &= s\dot{s} \\
&= s\left(-K\operatorname{sign}s + c^{\top}Bd\right) \\
&= -K|s| + c^{\top}Bds \\
&\leq -K|s| + \left|c^{\top}Bd\right||s| \\
&= -|s|\left(K - \left|c^{\top}Bd\right|\right).
\end{aligned}
$$

Since $K > \sup_{t\geq 0}|c^{\top}Bd(t)|$, the solution to above differential inequality converges to zero (i.e. $V = 0$) in finite-time. Hence, s becomes zero in finite-time, and sliding mode occurs in the system in finite-time. So, when sliding mode occurs in the system, we have $s(x) = c_1^{\top}x_1 + x_2 = 0$. This implies $x_2 = -c_1^{\top}x_1$. Using this in (1.12), we write sliding motion dynamics as

$$
\begin{aligned}
\dot{x}_1 &= \left(A_{11} - A_{12}c_1^{\top}\right)x_1 \\
x_2 &= -c_1^{\top}x_1.
\end{aligned}
$$

It is seen that during sliding mode, the system is completely independent of external disturbance; however, it depends on sliding surface parameter. Another important observation is that the order of the system is reduced by the number of control inputs. The essential steps in SMC are that (i) design the sliding surface such that sliding motion dynamics is stable and (ii) design the SMC law to bring sliding mode in the system in finite-time.

Until the sliding manifold is reached, the control signal is continuous. This is known as *reaching phase*. In this phase, the system is not robust since closed loop system is still perturbed by disturbances. However, as sliding mode occurs in finite-time, the system becomes robust against these (matched) disturbances. This sliding mode is often called as *sliding phase*.

Example 1.2 Consider the system (1.11) with

$$
A = \begin{bmatrix} -1 & 2 \\ 2 & 1 \end{bmatrix} \quad \text{and} \quad B = \begin{bmatrix} 2 \\ 1 \end{bmatrix}.
$$

First of all, we see that the pair (A, B) is controllable. So, we now proceed to design SMC for the above system. Since the above system is not in regular form, we transform the system into regular form using a nonsingular transformation

$$
T = \begin{bmatrix} 1 & -1 \\ 0 & 1 \end{bmatrix}.
$$

Then, defining the transformation $z = Tx$, the transformed system can be written as

$$\dot{z} = \begin{bmatrix} -5 & -10 \\ 2 & 5 \end{bmatrix} z + \begin{bmatrix} 0 \\ 1 \end{bmatrix} (u + d).$$

Assume that the disturbance d is upper bounded by one. Design the sliding variable $s = c^\top z = c_1 z_1 + z_2$ such that during sliding the dynamics is stable. That means, when $s = 0$, the following dynamics

$$\dot{z}_1 = -5z_1 - 10z_2$$
$$z_2 = -c_1 z_1$$

is stable. Thus, the value of c_1 must satisfy $c_1 < 0.5$, and it is selected as 0.3. Once the sliding surface is designed, the control law can now be given as

$$u = -\left(\begin{bmatrix} 0.5 & 1 \end{bmatrix} \begin{bmatrix} 0 \\ 1 \end{bmatrix} \right)^{-1} \left(\begin{bmatrix} 0.5 & 1 \end{bmatrix} \begin{bmatrix} -5 & -10 \\ 2 & 5 \end{bmatrix} z + 1.5 \mathrm{sign} s \right).$$

The simulation result is shown in Figs. 1.7 and 1.8. It is seen that the state trajectories converge to zero asymptotically with the SMC. Similar plot of state trajectory in the phase plane is also shown in Fig. 1.8. The sliding motion of the system in phase plane is started in finite-time.   □

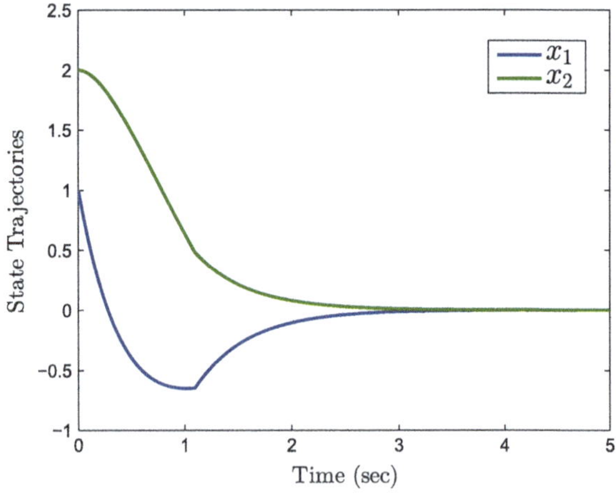

Fig. 1.7 Response of the system

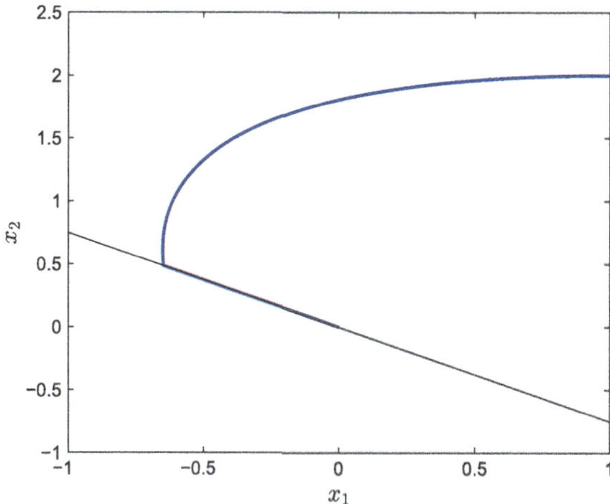

Fig. 1.8 State trajectory in phase plane

1.4 Discrete-Time Sliding Mode

In practical applications, nowadays almost all the controllers are implemented through digital processor or computers. For achieving almost continuous-time performance, the sampling period must be sufficiently small. However, it is not possible to choose sampling period smaller than some value due to computational and physical constraints. So, SMC is designed from the discrete-time plant rather than from continuous-time. The control is known as discrete-time sliding mode (DTSM) control.

Unlike the continuous-time SMC, η-reachability is not a sufficient condition for the existence of sliding mode for DTSM control. So, DTSM has drawn attention of many researchers in sliding mode community. The main focus here is to propose necessary and sufficient condition for the existence of sliding mode in discrete-time system. Many reaching conditions are developed for discrete-time system that ensures convergence of trajectory to sliding manifold and is often referred as *reaching law* for DTSM control. It is also to be noted that in continuous-time design the exact sliding mode is possible if the control is applied at infinite frequency. But, this is not achievable due to natural constraints. As a result of this, the trajectory remains bounded in the vicinity of sliding manifold and this bound is dependent on the factor that determines how fast control signal is applied. It is well known that the bound in DTSM depends on sampling interval, bounds of external disturbances and system uncertainties. The system trajectory will remain in the vicinity of sliding manifold and is known as *quasi-sliding mode* (QSM). The corresponding band is known as QSM band. Generally, constant sampling period is chosen in the design and analysis of DTSM control in almost all the works in literature.

Different reaching laws have been proposed that ensures convergence to discrete-time sliding manifold in finite number of steps. Broadly, there are two philosophies in the design of different reaching laws that have appeared in the literature. The one, that is motivated from continuous-time reaching condition, is switching-based reaching law. And the other one is switching-free reaching laws. In discrete-time systems, the sliding mode is possible even when the signum term does not appear in the control law. However, this is not true for the continuous-time systems.

1.4.1 Switching-Based Reaching Law

In this type of reaching laws, the control law contains the switching term as it is done in continuous-time case. Also, this reaching laws have been obtained from its continuous counter parts. The most popular among this is Gao's reaching law which is given below

$$s(k + 1) = (1 - q\tau)s(k) - \varepsilon\tau \operatorname{sign}s(k) + d(k) - d_0 - d_1 \operatorname{sign}s(k)$$

where τ is the constant sampling period at which control is applied to the plant, $\varepsilon > 0$, and q is such that $0 < 1 - q\tau < 1$. The disturbance $d(k)$ is bounded with known bounds $d_l \leq d(k) \leq d_u$. The mean and spread of the disturbance are defined using the constants d_l and d_u as

$$d_0 = \frac{d_u + d_l}{2} \quad \text{and} \quad d_1 = \frac{d_u - d_l}{2}.$$

The system trajectory always crosses and recrosses the sliding manifold subsequently once it reaches this manifold. The sufficient condition for such motion can be obtained by satisfying the relation

$$\operatorname{sign}s(k) = -\operatorname{sign}s(k + 1) = \operatorname{sign}s(k + 2).$$

Using this condition, it can be shown that the following is a sufficient condition for crossing and recrossing,

$$d_1 < \frac{q\tau\varepsilon\tau}{2(1 - q\tau)}.$$

So, if the disturbance bound satisfies the above condition, the sliding trajectory reaches the sliding manifold in finite number of discrete steps and remains ultimately bounded within a QSM band. This QSM band is given as

$$\left\{ x \in \mathbb{R}^n : |s| = |c^\top x| \leq 2d_1 + \varepsilon\tau \right\}.$$

Apart from this, there are different reaching laws that have been proposed in the literature to improve the steady-state performance. However, these are not reported here because of irrelevance to the topics covered in the following chapters.

1.4.2 Switching-Free Reaching Law

In this type of reaching laws, the switching term does not appear in the control expression. Nevertheless, the state trajectory is attracted towards the sliding manifold. This feature is not seen in the case of continuous-time system. One of the popular reaching laws is given by

$$s(k + 1) = 0.$$

This reaching law claims that if $s(k) \neq 0$, then it will reach zero in immediate next discrete step. So, in the initial step, in case $s(0)$ is not equal to zero, the control derived using this reaching would ensure $s(1) = 0$. As it forces the trajectory to reach the sliding manifold in one step, large control effort is required. However, in the presence of disturbance, the trajectory does not remain exactly on the manifold but within a band called QSM band. Note that in this type of reaching laws the crossing and recrossing condition is relaxed for the DTSM control. There is also another reaching law proposed by Bartoszewicz that allows the sliding variable go to zero in some finite number of desired steps which is discussed in Chap. 5 in detail.

It is to be noted in all these applications the control is implemented in discrete instants with constant sampling period. Though this technique is easy to analyse and design, it is not cost-effective in terms of control tasks are executed. So, aperiodic control implementation is desired with sampling interval as large as possible, provided stability is retained. However, it is difficult to ensure the stability of the closed loop system with aperiodic control implementation due to unavailability of many stability tools. In spite of this, many design techniques are available for aperiodic sampled-data system that guarantees the closed loop system stability.

1.5 Summary

In this chapter, introduction to computer-controlled system is briefly discussed. The configuration of such systems in digital platform is presented, and then, different control design methodologies available for analysing the stability are given. Some basic background on *periodic* and *aperiodic* control execution (in this case, Lebesgue sampling) is presented with motivation. In this context, the development of event-triggering strategy is given and some design techniques are also presented. SMC is also introduced in this chapter to familiarize the readers. The sliding motion and system stability are explained. Also, DTSM is briefly reviewed.

1.6　Notes and References

The basic introduction to computer-controlled systems can be found in [2, 3]. The different design approaches for a sample-data system are thoroughly presented in [3] and particularly the discrete-time analysis in [2]. Event-triggering-based control design is first presented in [4], and more detailed discussion can be found in [5–10]. The recent advances and applications of event-triggered control are reported in [11–13]. For the notion of ISS, the readers may refer to [14], and the detailed treatment of comparison function can be found in [15]. A formal introduction of VSS and SMC introduced here is available in [16–18]. Also, the readers may see [19, 20], [21]. The notion of solution during sliding mode is defined in the sense of Filippov [1]. For the design of DTSM control, we refer [22–31]. The Gao's reaching law presented here is given in [24].

References

1. A.F. Filippov, *Differential Equations With Discontinuous Right-Hand Sides* (Kluwer Academic Publishers, Dordrecht, 1988)
2. K.J. Åström, B. Wittenmark, *Computer-Controlled Systems: Theory and Design* (Prentice Hall, Upper Saddle River, 2002)
3. T. Chen, B.A. Francis, *Optimal Sampled-Data Control Systems*, Communications and Control Engineering (Springer, London, 1995)
4. K.-E. Årźen, A simple event-based PID controller, in *Proceedings of 14th IFAC World Congrress, Beijing, China* (1999), pp. 423–428
5. K.J. Åström, B. Bernhardsson, Comparison of Riemann and Lebesgue sampling for first order stochastic systems, in *Proceedings of 41st IEEE Conference on Decision and Control, Las Vegas, USA* (2002), pp. 2011–2016
6. P. Tabuada, Event-triggered real-time scheduling of stabilizing control tasks. IEEE Trans. Autom. Control **52**(9), 1680–1685 (2007)
7. Y.-K. Xu, X.-R. Cao, Lebesgue-sampling-based optimal control problems with time aggregation. IEEE Trans. Autom. Control **56**(5), 1097–1109 (2011)
8. W.P.M.H. Heemels, K.H. Johansson, P. Tabuada, An introduction to event-triggered and self-triggered control, in *Proceedings of 51st IEEE Conference of Decision and Control, Hawai, USA* (2010), pp. 3270–3285
9. P. Tabuada, X. Wang, Preliminary results on state-triggered stabilizing control tasks, in *Proceedings of 45th IEEE Conference on Decision and Control, San Deigo, USA* (2006), pp. 282–287
10. J. Lunze, D. Lehmann, A state-feedback approach to event-based control. Automatica **46**(1), 211–215 (2010)
11. D.P. Borger, W.P.M.H. Heemels, Event-separation properties of event-triggered control systems. IEEE Trans. Autom. Control **59**(10), 2644–2656 (2014)
12. A. Girard, Dynamic triggering mechanisms for event-triggered control. IEEE Trans. Autom. Control **60**(7), 1992–1997 (2015)
13. E. Garcia, P.J. Antsaklis, Model-based event-triggered control for systems with quantization and time-varying network delays. IEEE Trans. Autom. Control **58**(2), 422–434 (2013)
14. E.D. Sontag, Smooth stabilization implies coprime factorization. IEEE Trans. Autom. Control **34**(4), 435–443 (1989)

15. C.M. Kellett, A compendium of comparison function results. Math. Control Signals Syst. **26**(3), 339–374 (2014)
16. V.I. Utkin, Variable structure systems with sliding modes. IEEE Trans. Autom. Control **22**(2), 212–222 (1977)
17. V.I. Utkin, *Sliding Modes and Their Applications in Variable Structure Systems*, Translated from the Russian by A. Parnakh (MIR Publishers, Moscow, 1978)
18. V.I. Utkin, *Sliding Modes in Control and Optimization* (Springer, New York, 1992)
19. K.D. Young, V.I. Utkin, Ü. Özgüner, A control engineer's guide to sliding mode control. IEEE Trans. Control Syst. Technol. **7**(3), 328–342 (1999)
20. Y. Shtessel, C. Edwards, L. Fridman, A. Levant, *Sliding Mode Control and Observation* (Birkhäuser, Basel, 2014)
21. C. Edwards, S.K. Spurgeon, *Sliding Mode Control: Theory and Applications* (CRC Press, Taylor and Francis Group, 1998)
22. S.Z. Sarpturk, I. Istefanopulus, O. Kaynak, On the stability of discrete-time sliding mode control systems. IEEE Trans. Autom. Control **32**(10), 930–932 (1987)
23. K. Furuta, Sliding mode control of a discrete system. Syst. Control Lett. **14**(2), 145–152 (1990)
24. W. Gao, Y. Wang, A. Homaifa, Discrete-time variable structure control systems. IEEE Trans. Ind. Electron. **42**(2), 117–122 (1995)
25. G. Bartolini, A. Ferrara, V.I. Utkin, Adaptive sliding mode control in discrete-time systems. Automatica **31**(5), 769–773 (1995)
26. A. Bartoszewicz, Discrete-time quasi-sliding-mode control strategies. IEEE Trans. Ind. Electron. **45**(4), 633–637 (1998)
27. A. Bartoszewicz, P. Lesniewski, New switching and nonswitching type reaching laws for SMC of discrete time systems. IEEE Trans. Control Syst. Technol. **24**(2), 670–677 (2016)
28. W.-C. Su, S.V. Drakunov, Ü. Özgüner, An $O(T^2)$ boundary layer in sliding mode for sampled-data systems. IEEE Trans. Autom. Control **45**(3), 482–485 (2000)
29. M.L. Corradini, G. Orlando, Variable structure control of discretized continuous-time systems. IEEE Trans. Autom. Control **43**(9), 1329–1334 (1998)
30. G. Golo, Č. Milosavljević, Robust discrete-time chattering free sliding mode control. Syst. Control Lett. **41**(1), 19–28 (2000)
31. S. Chakrabarty, B. Bandyopadhyay, A generalized reaching law for discrete time sliding mode control. Automatica **52**, 83–86 (2015)

Chapter 2
Event-Triggered Sliding Mode Control for Linear Systems

SMC is a robust control technique that ensures the robust (insensitive) stabilization of the system against (matched) disturbance. This property holds when it is designed and implemented in continuous-time domain. On the other hand, in digital implementation of SMC the system trajectory slides in the vicinity of sliding manifold. Many techniques and design methodologies have appeared in the literature for reducing the sliding band and improving the accuracy. Nevertheless, a significant improved performance can be obtained using some novel control implementation strategy, namely event-triggering technique.

This chapter presents the design of event-triggered SMC for LTI system. It is seen that while implementing the control in event-triggered manner the system trajectory remains bounded in the vicinity of sliding manifold. However, the bound of sliding trajectory depends only on some design parameter. This is defined as *practical sliding mode*. The proposed triggering rule ensures the sufficient condition for the existence of *practical sliding mode*, and hence by designing a suitable parameter, any desired accuracy can be achieved. Further, in order to show the stability of the event-triggered system, a positive lower bound between any two consecutive triggering instants is established. However, the global stability is not achieved with this triggering rule. So, this chapter also presents a new event-triggering rule that ensures global stability of the LTI systems. At the end, the event-triggered SMC is developed for MIMO LTI systems. Numerical examples are presented to illustrate the design and performance of event-triggered SMC.

2.1 System Description

Consider the dynamical system

$$\dot{x} = Ax + B(u + d)$$

© Springer International Publishing AG, part of Springer Nature 2018
B. Bandyopadhyay and A. K. Behera, *Event-Triggered Sliding Mode Control*, Studies
in Systems, Decision and Control 139, https://doi.org/10.1007/978-3-319-74219-9_2

as in (1.11) with matched disturbance. Assumption 1.1 also holds on the disturbance d. Before designing the event-triggered SMC, we recall the continuous-time SMC designed for this system.

The sliding variable is chosen as $s = c^\top x$, and the sliding manifold is defined as given in (1.14). To see the design of sliding surface parameter c, refer to Sect. 1.3.2. The control law that brings the sliding mode in the system is given in (1.15) and is rewritten here

$$u = -(c^\top B)^{-1} \left(c^\top A x + K \operatorname{sign} s \right).$$

In practical applications, the control law (1.15) is applied using digital processors. So, the digital control is obtained by simply replacing the continuous states by the discrete ones in the continuous-time control (1.15) expression and then it is implemented. This control is held constant until the next sampling/triggering instant and is given below as

$$u(t) = -(c^\top B)^{-1} \left(c^\top A x(t_i) + K \operatorname{sign} s(t_i) \right) \tag{2.1}$$

for all $t \in [t_i, t_{i+1})$ and $i \in \mathbb{Z}_{\geq 0}$. In this chapter, we discuss the stability of the system (1.11) with the discrete control (2.1) when it is to be implemented in event-triggered framework. Let $\{t_i\}_{i=0}^\infty$ be a sequence of triggering instant generated at which the control signal is applied to the plant. We define $T_i := t_{i+1} - t_i$ as inter-event time of the triggering rule. Due to the discrete implementation (also in event-triggering) of the controller, the error $e(t) = x(t_i) - x(t)$ is introduced in the system for all time $t \in [t_i, t_{i+1})$. Note that if the control is applied in continuous manner then the error $e(t) = 0$ for all time.

It is very well known that when SMC is implemented in digital platform, (exact) sliding mode is not possible due to discrete nature of the control law. However, the system trajectory remains bounded in the vicinity of sliding manifold (1.14). If the control is implemented at some constant sampling period, then the bound of the sliding trajectory also depends on this sampling period. Also, it depends on the maximum bound of the disturbance. In other words, steady-state bounds of system trajectory depend on the chosen sampling period and the disturbance bound.

In this chapter, event-triggered implementation of SMC is presented that improves steady-state performance of the system. The robustness of the system against matched disturbances is analysed here to address the robust performance of the system. So, the next sections of this chapter present the event-triggered design of SMC and the closed loop system stability.

2.2   Event-Triggered Sliding Mode Control

In event-triggered SMC, the steady-state bound is independent of sampling interval and also other plant parameters. The event parameter completely determines the steady-state bound during sliding mode in the system in the presence of disturbance.

Due to this property, any desired steady-state bounds can be achieved using event-triggered SMC. We define such motions in the system as *practical sliding mode* and the corresponding band as *practical sliding mode band*.

The notation $x(t, x(t_0))$ denotes the trajectory of the system (1.11) at time t with initial condition $x(t_0)$. However, in most of the chapters, it is denoted by simply $x(t)$ without any initial condition if no confusion arises.

Definition 2.1 (*Practical Sliding Mode*) Consider any trajectory $x(t)$ of the system (1.11) and sliding manifold (1.14) for any stable sliding function $s(x(t))$. The system is said to be in *practical sliding mode* if given any $\varepsilon > 0$ there exists a $t_1 \in [t_0, \infty)$ such that the sliding trajectories remain bounded in the vicinity of sliding manifold with the bound given by ε for all time $t \geq t_1$. That means, $\|s(t)\| \leq \varepsilon$ for all time $t \geq t_1$. The corresponding bound of sliding trajectory is called as *practical sliding mode band*.

In the below, some relevant definitions are introduced to characterize the event-triggering mechanism for an ETCS in terms of the inter-event time.

Definition 2.2 An event-triggered system with SMC is said to have a robust global event property if

$$\inf_{\substack{x \in \mathbb{R}^n \\ i \in \mathbb{Z}_{\geq 0} \\ |d(t)| \leq d_0}} T_i > a_T$$

for some positive constant a_T.

Definition 2.3 An event-triggered system with SMC is said to have a robust semi-global event property if for every subset $\mathscr{D}$ in $\mathbb{R}^n$,

$$\inf_{\substack{x \in \mathscr{D} \subset \mathbb{R}^n \\ i \in \mathbb{Z}_{\geq 0} \\ |d(t)| \leq d_0}} T_i > a_T$$

for some positive constant a_T.

Definition 2.4 An event-triggered system with SMC is said to have a robust local event property if for some subset $\mathscr{D}$ in $\mathscr{X} \subset \mathbb{R}^n$,

$$\inf_{\substack{x \in \mathscr{D} \cap \mathscr{X}, \mathscr{D} \subset \mathscr{X} \subset \mathbb{R}^n \\ i \in \mathbb{Z}_{\geq 0} \\ |d(t)| \leq d_0}} T_i > a_T$$

for some positive constant a_T.

For the sake of brevity, we call the robust global event property (respectively, robust semi-global and robust local event properties) simply as global property (respectively, semi-global property and local property) in the following discussions. From the above notions, it implies that if the event-triggered system has global property, then it also has semi-globally property, and similarly, semi-global property implies the local property for an ETCS. But, the statements do not hold in reverse direction.

In event-triggered SMC, the main objective is to design the switching gain, K, such that the practical sliding mode is enforced in the system. We will see that this gain condition for the practical sliding mode is not same as that of continuous-time sliding mode case. Following sufficient conditions enable to design first the steady-state bound of sliding trajectory and then develop the event-triggering scheme to achieve this objective.

Theorem 2.1 *Consider the system (1.11) and the control law (2.1). Let $\alpha > 0$ be given such that*

$$\|c\|\,\|A\|\,\|e(t)\| < \alpha \tag{2.2}$$

holds for all $t \geq 0$. Then, the practical sliding mode is enforced in the system with the control law (2.1) if

$$K > \sup_{t \geq 0} \left|c^\top B d(t)\right| + \alpha. \tag{2.3}$$

Proof To prove the existence of practical sliding mode, consider the Lyapunov function $V = \frac{1}{2}s^2$. We show that for all time $t \in [t_i, t_{i+1})$ and $i \in \mathbb{Z}_{\geq 0}$, the sliding trajectories are brought to vicinity of $\mathscr{S}$, as defined in (1.14), by the control law (2.1). Now, differentiating V with respect to time along the system trajectories, we obtain

$$\begin{aligned}
\dot{V}(s(t)) &= s(t)\dot{s}(t) \\
&= s(t)\left(c^\top A x(t) - c^\top A x(t_i) - K\mathrm{sign}(s(t_i)) + c^\top B d(t)\right) \\
&= -s(t)\left(c^\top A e(t) + K\mathrm{sign}(s(t_i)) - c^\top B d(t)\right) \\
&\leq |s(t)|\left|c^\top A e(t)\right| - s(t)K\mathrm{sign}(s(t_i)) + |s(t)|\left|c^\top B\right|d_0.
\end{aligned} \tag{2.4}$$

It is to be noted that the sign of sliding variable, s, does not change until it reaches manifold $\mathscr{S}$. So, it can be concluded that $\mathrm{sign}(s(t_i)) = \mathrm{sign}(s(t))$. Now, using this, (2.2) and (2.3) in (2.4), it yields

$$\begin{aligned}
\dot{V}(s(t)) &\leq -|s(t)|\left(K - \alpha - \left|c^\top B\right|d_0\right) \\
&\leq -\eta|s(t)|
\end{aligned} \tag{2.5}$$

for some $\eta > 0$. That means as long as $\mathrm{sign}(s(t_i)) = \mathrm{sign}(s(t))$ is satisfied the sliding trajectories are attracted towards the sliding manifold $\mathscr{S}$ due to

$$\dot{V}(s) \leq -\eta|s|.$$

This ensures that the sliding trajectory converges towards the sliding manifold. When the trajectory hits the manifold $s = 0$, the sign of s changes in the time interval $[t_i, t_{i+1})$ for some $i < +\infty$ and $\dot{V} < 0$ cannot be guaranteed for all time. However, the trajectory crosses the manifold and eventually it moves away from this due to $\dot{s}(t) < -\eta \operatorname{sign} s(t_i)$ in that corresponding triggering interval. However, the system trajectory cannot increase beyond a certain value due to relation (2.2). This can be proved as follows.

We obtain the maximum value of the bound for s such that the sliding trajectory remains bounded within it. Since the relation (2.2) does not allow the trajectory to go beyond a certain value, it is natural to calculate the maximum deviation of sliding trajectory within a triggering interval around $\mathscr{S}$.

An estimate of sliding band can be given as follows. Now, we let $\|c\| \|A\| \|e(t)\| < \alpha$ holds for all time. Using this in the following relation, we obtain

$$
\begin{aligned}
|s(t_i) - s(t)| &= \left| c^\top x(t_i) - c^\top x(t) \right| \\
&\leq \|c\| \|e(t)\| \\
&< \frac{\alpha}{\|A\|}.
\end{aligned}
\tag{2.6}
$$

This gives the maximum deviation of the sliding trajectory from its immediate past sampled value. From the above relation, it is seen that if the maximum deviation of sliding trajectory in one triggering interval is more than $\alpha \|A\|^{-1}$, then a triggering instant is generated. Otherwise, the relation (2.2) may not hold. Thus, the sliding trajectory decreases by an amount $\alpha \|A\|^{-1}$ from its value at the previous triggering instant. In turn, the maximum value of sliding band can be obtained by substituting $s(t_i) = 0$ as $\alpha \|A\|^{-1}$. Therefore, the system trajectories are always bounded within the region given as

$$
\left\{ x \in \mathbb{R}^n : |s| = \left| c^\top x \right| < \frac{\alpha}{\|A\|} \right\}.
$$

Therefore, the sliding trajectory remains bounded within a band that is dependent only on the design parameter α. This guarantees the existence of practical sliding mode in the system, and thus, the proof is completed. $\qquad\square$

Remark 2.1 The value of α determines the steady-state bound of system trajectories; hence, it must be chosen to satisfy the design constraints. It is always possible to provide the upper bound on α that guarantees satisfactory performance of the system. However, for large values of α, large steady-state bound is obtained with minimum computational burden.

Remark 2.2 For a single integrator $\dot{x} = u + d$, the sliding variable can be taken as $s = x$. The SMC for the system can be given by $u = -K \operatorname{sign} x$ with $K > d_0$. The practical sliding mode band in this case can be obtained as $\{x \in \mathbb{R} : |x| \leq \alpha\}$. However, the triggering instant is generated by the triggering rule given by $t_{i+1} = \inf\{t > t_i : |e(t)| \geq \alpha\}$.

2.2.1 Stability of Sliding Motion

Here, the stability of the system during (practical) sliding mode is discussed. Consider the regular form of the system (1.11) as it is given by (1.12)–(1.13). This is rewritten below again,

$$\dot{x}_1 = A_{11}x_1 + A_{12}x_2$$
$$\dot{x}_2 = A_{21}x_1 + A_{22}x_2 + B_2(u + d).$$

The sliding variable coefficient is written as $c = [c_1^\top \ 1]^\top$. In the event-triggered sliding mode, the sliding variable does not go to zero but is bounded. So, during this sliding mode, it can be written as

$$x_2 = -c_1^\top x_1 + s.$$

The system dynamics is obtained by substituting the above relation in (1.12). This becomes a reduced order system and is given as

$$\dot{x}_1 = \left(A_{11} - A_{12}c_1^\top\right)x_1 + A_{12}s \tag{2.7}$$
$$x_2 = -c_1^\top x_1 + s. \tag{2.8}$$

The closed loop system stability can be given in the following proposition.

Proposition 2.1 *Consider the systems (2.7). Let the matrix* $A_c = A_{11} - A_{12}c_1^\top$ *has all its eigenvalues with negative real parts. Then, the system trajectory remains bounded with an ultimate bound*

$$\mathscr{B}_1 := \left\{x_1 \in \mathbb{R}^{n-1} : \|x_1\| \leq \Theta\right\} \tag{2.9}$$

where $\Theta = \frac{2\alpha\|PA_{12}\|}{\lambda_{\min}\{Q\}\|A\|}.$

Proof Consider the Lyapunov function $V_1 = x_1^\top P x_1$ where $P = P^\top > 0$. Then, differentiating V_1,

$$\dot{V}_1 = \dot{x}_1^\top P x_1 + x_1^\top P \dot{x}_1$$
$$= x_1^\top (A_c^\top P + P A_c)x_1 + 2x_1^\top P A_{12}s. \tag{2.10}$$

Recall that given any Hurwitz matrix A_c and $Q = Q^\top > 0$ there exists a $P > 0$ such that $A_c^\top P + P A_c = -Q$. Using this,

$$\dot{V}_1 = -x_1^\top Q x_1 + 2x_1^\top P A_{12}s$$
$$\leq -\lambda_{\min}\{Q\}\|x_1\|^2 + 2\|x_1\|\|P A_{12}\||s|.$$

Now, applying Young's inequality[1] to the second term in the above relation (with $a = 2\|PA_{12}\||s|$, $b = \|x_1\|$, and $\varepsilon = \lambda_{\min}\{Q\}$), it yields

$$\dot{V}_1 \leq -\frac{\lambda_{\min}\{Q\}}{2}\|x_1\|^2 + 2\frac{\|PA_{12}\|^2}{\lambda_{\min}\{Q\}}|s|^2.$$

From Theorem 2.1, we have $|s| < \frac{\alpha}{\|A\|}$. This implies

$$\dot{V}_1 \leq -\frac{\lambda_{\min}\{Q\}}{2}\|x_1\|^2 + \frac{\lambda_{\min}\{Q\}}{2}\Theta^2 \tag{2.11}$$

where Θ is defined in (2.9). It is, thus, concluded that if $\|x_1\| > \Theta$, the system trajectories are attracted to $\mathscr{B}_1$ as $\dot{V}_1 < 0$. On the other hand, the trajectories remain bounded within the ball $\mathscr{B}_1$ whenever $\|x_1\| \leq \Theta$. This implies $\mathscr{B}_1$ is attractive, and hence, the trajectories are ultimately bounded. This completes the proof. $\square$

Remark 2.3 From this proposition, it is clear that system trajectories of sliding mode dynamics is bounded, i.e. $x_1 \in \mathscr{B}_1$. As a result, the system trajectories also remain bounded in view of relation (2.8).

2.2.2 Stability of Event-Triggered System

Practical sliding mode is guaranteed in the system if the relation (2.2) holds. So, the relation (2.2) is used for designing event-triggering rule for SMC. Triggering rule is generally developed such that the closed loop system is stable. In this case, it is designed such that the practical sliding mode exists and ensures stable sliding motion dynamics. The event-triggering rule is

$$t_{i+1} = \inf\left\{t > t_i : \|c\|\|A\|\|e(t)\| \geq \sigma\alpha\right\} \tag{2.12}$$

for some $\sigma \in (0, 1)$. Here, the constant σ is introduced to account for some unavoidable delays in the control implementation. The triggering instants are generated whenever (2.12) is violated. This leads to $\|c\|\|A\|\|e(t)\| < \alpha$ for all time $t \geq 0$. Again, the condition (2.2) is also true. Here, in this chapter, it is assumed that at the triggering instant the control signal is updated.

In the next, the stability of ETCS with the proposed triggering rule is discussed. Let $\{t_i\}_{i=0}^{\infty}$ be the sequence of triggering instants generated due to (2.12). Since this

---

[1] We use here Young's inequality for exponent two that states for any nonnegative real numbers a, b and every $\varepsilon > 0$, the following holds,

$$ab \leq \frac{a^2}{2\varepsilon} + \frac{\varepsilon b^2}{2}.$$

triggering sequence assigns the control tasks to the processor, it is imperative to ensure that the inter-event time—time interval between any two consecutive triggering instants—is always lower bounded by some finite positive quantity.

Theorem 2.2 *Consider the system (1.11) and the control law (2.1). Let $\{t_i\}_{i=0}^{\infty}$ be the triggering sequences generated by (2.2) and $\sigma \in (0, 1)$ be given. Then, the inter-event time, defined by $T_i := t_{i+1} - t_i$, satisfies*

$$T_i \geq \frac{1}{\|A\|} \ln \left(1 + \sigma \frac{\alpha}{\|c\|(\rho(\|x(t_i)\|) + \beta)} \right) =: \Xi(x(t_i), \sigma) \tag{2.13}$$

where β and the real-valued function $\rho(\|x(t_i)\|) : \mathbb{R}^n \to \mathbb{R}_{\geq 0}$ are given as

$$\rho(\|x(t_i)\|) = \left\| A - B(c^\top B)^{-1} c^\top A \right\| \|x(t_i)\| \tag{2.14}$$

and

$$\beta = \left\| B(c^\top B)^{-1} \right\| K + \|B\|d_0. \tag{2.15}$$

Proof Define $\Gamma = \{t \in [t_i, t_{i+1}) : \|e(t)\| = 0\}$. The proof follows by finding the maximum growth of $\|e(t)\|$ during any triggering interval and then establish that there exists a minimum positive lower bound for T_i. So, differentiating error $\|e(t)\|$ for all $t \in [t_i, t_{i+1})\backslash\Gamma$

$$
\begin{aligned}
\frac{\mathrm{d}}{\mathrm{d}t}\|e(t)\| &\leq \left\| \frac{\mathrm{d}}{\mathrm{d}t}e(t) \right\| = \left\| \frac{\mathrm{d}}{\mathrm{d}t}x(t) \right\| \\
&= \left\| Ax(t) - B(c^\top B)^{-1} c^\top Ax(t_i) - B(c^\top B)^{-1} K \operatorname{signs}(t_i) + Bd(t) \right\| \\
&= \left\| Ax(t_i) - Ae(t) - B(c^\top B)^{-1} c^\top Ax(t_i) - B(c^\top B)^{-1} K \operatorname{signs}(t_i) + Bd(t) \right\| \\
&\leq \|A\|\|e(t)\| + \left\| \left(A - B(c^\top B)^{-1} c^\top A \right) x(t_i) \right\| + \|B(c^\top B)^{-1} K \| + \|B\|d_0 \\
&= \|A\|\|e(t)\| + \rho(\|x(t_i)\|) + \beta. \tag{2.16}
\end{aligned}
$$

The solution to the differential inequality (2.16) can be obtained using comparison Lemma [1]. Thus, the solution to (2.16) with the initial condition $\|e(t_i)\| = 0$ is obtained as

$$\|e(t)\| \leq \frac{\rho(\|x(t_i)\|) + \beta}{\|A\|} \left(e^{\|A\|(t-t_i)} - 1 \right) \tag{2.17}$$

for all time $t \in [t_i, t_{i+1})$. Again, at the instant of triggering, $\|c\|\|A\|\|e(t_{i+1})\| = \sigma\alpha$, so

$$\frac{\sigma\alpha}{\|c\|\|A\|} = \|e(t_{i+1})\| \leq \frac{\rho(\|x(t_i)\|) + \beta}{\|A\|}\left(e^{\|A\|T_i} - 1\right). \tag{2.18}$$

The lower bound for T_i follows immediately from the rearrangement of (2.18) as given in (2.13). It is to be noted that this lower bound $\Xi(x(t_i), \sigma)$ is always a finite positive quantity for all finite values of states. In other words, $\Xi(x(t_i), \sigma) > 0$ for all time $t \in \mathbb{R}_{\geq 0}$. Indeed, we have $t_{i+1} > t_i + a_T$ for some $a_T > 0$ and thus it ensures no Zeno phenomenon triggering sequence exists. Hence, the proof is completed. $\square$

2.2.3 Simulation Results

In this section, simulation results are shown for event-triggered SMC for a SISO LTI system. Consider an LTI system as

$$\dot{x} = \begin{bmatrix} 0 & 1 \\ 4 & 5 \end{bmatrix} x + \begin{bmatrix} 0 \\ 1 \end{bmatrix} (u + d).$$

The sliding surface is designed for this system as $s = [0.5\ 1]x$ such that the reduced order dynamics during sliding motion is stable. To see this, the reduced dynamics is given by

$$\dot{x}_1 = -0.5x_1$$

which is found to be asymptotically stable. Here, the disturbance signal $d(t)$ is bounded and the bound is selected as 0.5. For simulation purpose, we select $d(t) = 0.5\sin(10t)$. To design event-triggered SMC, the different event parameters are chosen as $\sigma = 0.85$ and $\alpha = 0.8$. So, the gain K of SMC must satisfy $K > 1.3$ according to (2.3). In simulation, it is designed as $K = 1.4$ that satisfies the design specifications. So, the event-triggered SMC is given as

$$u(t) = -\begin{bmatrix} 4 & 5.5 \end{bmatrix} x(t_i) - 1.4\,\mathrm{sign}s(t_i)$$

for all time $t \in [t_i, t_{i+1})$. The simulation is run in MATLAB with initial condition $x(0) = [5\ 2]^\top$ and $t_0 = 0$. The performance of event-triggered SMC is shown in Figs. 2.1 and 2.2.

The plot of state trajectory is shown in Fig. 2.1a. After some finite-time, the trajectory reaches the sliding manifold and remains bounded in a predesigned band for all time. As the time progresses, the sliding trajectory moves towards the equilibrium point with magnitude bounded by practical sliding mode band. The triggering instants are generated by the triggering rule at which the control signal is updated. The plot of inter-event time is shown in Fig. 2.1b. The inter-event time is lower bounded by some positive constant around 0.03. The time interval rises to approximately around

Fig. 2.1 Performance of event-triggered SMC for LTI systems

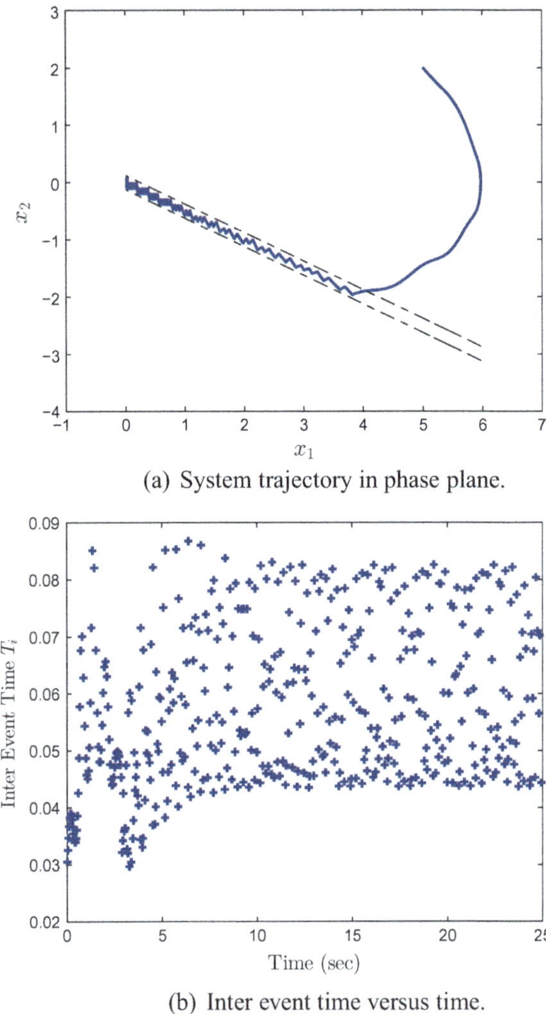

(a) System trajectory in phase plane.

(b) Inter event time versus time.

0.85. The inter-event times are bounded and it varies such that the sliding trajectory remains bounded by $\alpha \|A\|^{-1} = 0.124$.

The control signal is plotted in Fig. 2.2. The event-triggered control is held constant between two triggering instants. Once (practical) sliding mode is enforced, the control law switches which is also shown in Fig. 2.2. It is seen that the event-triggered SMC improves the system performance.

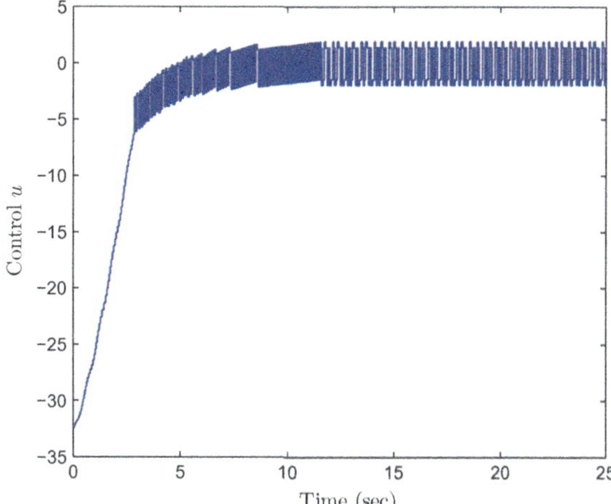

Fig. 2.2 Event-triggered SMC signal for LTI systems

2.3 Global Event-Triggered Sliding Mode Control

In the previous section, event-triggered realization of SMC is formulated and the corresponding triggering rule is also stated. In addition to this, the stability of event-triggered system is shown by establishing a positive lower bound for inter-event time. Here, it is shown that the triggering rule does not have a global property and hence only ensures the semi-global stability of the system. So, the main objective is to develop a triggering rule that guarantees the stability of the ETCS globally in the state space.

In ETCS, the triggering sequence generated by some triggering rule must be unbounded and strictly increasing. In other words, a positive lower bound for the inter-event time is guaranteed. Otherwise, an undesirable behaviour of the triggering sequence may occur which is known as *Zeno phenomenon* of triggering sequence. The main purpose of this section is to show triggering sequence generated by the proposed triggering rule has global property, i.e. it is *Zeno*-free globally.

From Definitions 2.3, it can be seen that the triggering rule (2.12) has semi-globally property. This can be easily derived from the fact that $\Xi(x(t_i), \sigma) > 0$ for every finite $x \in \mathscr{D}$, where $\mathscr{D}$ is any subset of $\mathbb{R}^n$. However, if $\|x\|$ goes to infinity then T_i tends to zero. In the following, a triggering scheme is developed with global event property for global robust stabilization of ETCS.

2.3.1 Global Event-Triggering Rule

It is concluded from the previous section that the triggering scheme $\|e\| < \frac{\sigma\alpha}{\|c\|\|A\|}$ ensures only semi-global stability. We introduce here a triggering scheme which gives global stability by modifying the triggering rule. The global event-triggering scheme is developed here as

$$t_{i+1} = \inf\left\{t > t_i : \|c\|\|A\|\|e(t)\| \geq \sigma(\|x(t_i)\| + \alpha)\right\} \tag{2.19}$$

for any given $\alpha \in (0, \infty)$ and $\sigma \in (0, 1)$. This triggering scheme implies that $\|c\|\|A\|\|e\| < \sigma(\|x(t_i)\| + \alpha)$ for all time. Due to this new triggering scheme, the steady-state bounds are increased compared to that of semi-global event-triggered SMC. Nevertheless, the system trajectories remain bounded with this global triggering scheme within a sliding band in the vicinity of sliding manifold (1.14). The size of this sliding band reduces as state decreases and attains a minimum value at the origin.

2.3.2 Design of Sliding Mode Control

In event-triggered SMC, the switching gain of SMC is always designed by consideration of event parameters. Semi-global event-triggering rule depends only on the constant parameter, so the switching gain in SMC is designed from this constant event-parameter only. In a similar manner, the state-dependent switching gain of SMC is designed in case of global event-triggering scheme since the triggering rule depends on the sampled state $x(t_i)$.

Global event-triggered SMC is given in the below as

$$u(t) = -(c^\top B)^{-1}\left(c^\top A x(t_i) + K(x(t_i))\mathrm{sign}s(t_i)\right) \tag{2.20}$$

for all time $t \in [t_i, t_{i+1})$. The switching gain, $K(x(t_i))$, depends on $x(t_i)$, and it satisfies the following properties:

a. $K(x(t_i)) > \sup_{t \geq 0}\left|c^\top B d(t)\right|,$
b. $K(x(t_i)) > \|x(t_i)\| + \alpha$ for all $x \in \mathbb{R}^n$.

The function $K(x(t_i))$ can be designed such that a and b are satisfied. Here, one of the simplest forms of such function is given by $K(x(t_i)) = K_1 + K_2\|x(t_i)\|$, where the gains K_1 and K_2 satisfy $K_1 > \left|c^\top B\right| d_0 + \alpha$ and $K_2 > 1$, respectively. It can be easily seen that the function $K(x(t_i))$ satisfies a. and b. as shown above.

The objective is to show the sliding trajectory reaches sliding manifold and moves towards the origin if the SMC (2.20) is designed as discussed above.

Theorem 2.3 *Consider the system (1.11) and the control law (2.20). Let $\alpha \in (0, \infty)$ be given such that*

$$\|c\|\|A\|\|e(t)\| < \|x(t_i)\| + \alpha \tag{2.21}$$

for all time $t \geq 0$ with t_0 being zero. Then, the control law (2.20) guarantees the sliding trajectory to remain within a band

$$\left\{ x \in \mathbb{R}^n : \left| c^\top x \right| \leq \frac{\alpha + \|x(t_i)\|}{\|A\|} \right\} \tag{2.22}$$

if $K(x(t_i))$ is designed as

$$K(x(t_i)) > \sup_{t \geq 0} \left| c^\top B d(t) \right| + \|x(t_i)\| + \alpha. \tag{2.23}$$

Proof Consider the Lyapunov function $V = \frac{1}{2}s^2$. Differentiating V with respect to time and using (2.20), one obtains

$$
\begin{aligned}
\dot{V}(s(t)) &= s(t)\dot{s}(t) \\
&= s(t)\left(c^\top A x(t) + c^\top B u(t_i) + c^\top B d(t)\right) \\
&= s(t)\left(-c^\top A e(t) - K_1 \operatorname{sign} s(t_i) - K_2 \|x(t_i)\| \operatorname{sign} s(t_i) + c^\top B d(t)\right).
\end{aligned}
\tag{2.24}
$$

In the last step of above relation, $e(t) = x(t_i) - x(t)$ is used. This can be further reduced using (2.21) to

$$
\begin{aligned}
\dot{V}(s(t)) &= -s(t)c^\top A e(t) - s(t)K_1 \operatorname{sign} s(t_i) - s(t)K_2 \|x(t_i)\| \operatorname{sign} s(t_i) + s(t)c^\top B d(t) \\
&\leq |s(t)| \left| c^\top A e(t) \right| - s(t)K_1 \operatorname{sign} s(t_i) - s(t)K_2 \|x(t_i)\| \operatorname{sign} s(t_i) \\
&\quad + |s(t)| \left| c^\top B d(t) \right| \\
&< |s(t)| \|x(t_i)\| + |s(t)|\alpha - s(t)K_1 \operatorname{sign} s(t_i) - s(t)K_2 \|x(t_i)\| \operatorname{sign} s(t_i) \\
&\quad + |s(t)| \left| c^\top B \right| d_0.
\end{aligned}
\tag{2.25}
$$

Note that until the system trajectory reaches the sliding manifold, $\operatorname{sign} s(t_i) = \operatorname{sign} s(t)$. Using this in the above,

$$
\begin{aligned}
\dot{V}(s(t)) &< -|s(t)|K_1 - |s(t)|K_2 \|x(t_i)\| + |s(t)| \|x(t_i)\| + |s(t)|\alpha + |s(t)| \left| c^\top B \right| d_0 \\
&= -|s(t)| \left(K_1 - \alpha - \left| c^\top B \right| d_0\right) - |s(t)| \left(K_2 - 1\right) \|x(t_i)\| \\
&< -|s(t)| \left(K_1 - \alpha - \left| c^\top B \right| d_0\right) \\
&\leq -\eta |s(t)|
\end{aligned}
\tag{2.26}
$$

for some $\eta > 0$. This establishes that the sliding manifold is always attractive whenever control signal is updated. However, when the trajectory is close to the vicinity of $s = 0$ then the trajectories crosses the manifold (1.14) due to unavailability of updated control signal. As a result of this, the sign of s changes in the time interval $[t_i, t_{i+1})$ for some $i < \infty$ and $\dot{V} < 0$ cannot be guaranteed. It is to be noted that the system trajectory cannot increase beyond a certain value due to (2.21). This can be proved as follows.

As the trajectory reaches sliding manifold, it crosses this manifold if control signal is not updated. So, the trajectory moves away from the manifold and in the mean time error in the event-triggered block also increases. This triggers the event at some time instant, and control is updated. This control signal forces the trajectory to move towards the sliding manifold again. This process continues while the trajectory moving towards the origin. So, the sliding band can be obtained by obtaining the maximum deviation of sliding trajectory in one triggering interval.

This can be derived as follows. We know that

$$
\begin{aligned}
|s(t_i) - s(t)| = \left| c^\top x(t_i) - c^\top x(t) \right| \\
\leq \|c\| \|e(t)\| \\
\leq \frac{\alpha + \|x(t_i)\|}{\|A\|}.
\end{aligned}
\tag{2.27}
$$

So, in one triggering interval the sliding trajectory moves at most by $\frac{\alpha + \|x(t_i)\|}{\|A\|}$ from its immediate past sampled value. From this, the maximum value of sliding band can be obtained by setting $s(t_i) = 0$ and is given in (2.22). Thus, the proof is completed. $\qquad\square$

In this global event-triggering scheme, the sliding band is state dependent and also it is more than the size of practical sliding mode band. This results at the cost of achieving global stability of event-triggering scheme. However, in practice a trade-off between $\|x(t_i)\|$ and α can be done for optimum bound on steady-state trajectory. For example, α may be decreased to some lower values for smaller size of sliding band and vice versa.

Like semi-global triggering scheme, the stability of sliding dynamics can also be shown by considering the reduced dynamics. Consider the system (1.12)–(1.13). When the system trajectory crosses the sliding manifold, it remains bounded by the band (2.22). Thus, the dynamics that governs the system is given as

$$
\dot{x}_1 = \left(A_{11} - A_{12} c_1^\top \right) x_1 + A_{12} s
\tag{2.28}
$$

$$
x_2 = -c_1^\top x_1 + s.
\tag{2.29}
$$

From the above dynamics and algebraic relations, it is concluded that if x_1 is bounded, x_2 also remains bounded since sliding variable s is constrained to be within a band. The stability is shown below.

Proposition 2.2 *Consider the dynamics (2.28) and (2.29). Let the matrix $A_c = A_{11} - A_{12} c_1^\top$ has all its eigenvalues with negative real parts. Then, the trajectories of the system remain bounded in the region given by*

$$
\mathscr{B}_2 := \left\{ x_1 \in \mathbb{R}^{n-1} : \|x_1\| \leq \frac{2\|PA_{12}\|(\alpha + \|x(t_i)\|)}{\lambda_{\min}\{Q\}\|A\|} \right\}
\tag{2.30}
$$

where P and Q are positive-definite matrices that satisfy $A_c^\top P + P A_c + Q = 0$.

Proof We consider the Lyapunov function $V_1 = x_1^\top P x_1$ with $P > 0$ satisfying above relation. Taking the time derivative of V_1 along system trajectories of (2.28),

$$\begin{aligned}
\dot{V}_1 &= \dot{x}_1^\top P x_1 + x_1^\top P \dot{x}_1 \\
&= -x_1^\top Q x_1 + 2 x_1^\top P A_{12} s \\
&\leq -\lambda_{\min}\{Q\}\|x_1\|^2 + 2\|x_1\|\|P A_{12}\||s|.
\end{aligned} \tag{2.31}$$

From Theorem 2.3, the sliding variable is bounded within (2.22). So, applying Young's inequality and then using this bound,

$$\dot{V}_1 \leq -\frac{\lambda_{\min}\{Q\}}{2}\|x_1\|^2 + \frac{\lambda_{\min}\{Q\}}{2}\Theta_1^2$$

where $\Theta_1 = \frac{2\|P A_{12}\|(\alpha+\|x(t_i)\|)}{\lambda_{\min}\{Q\}\|A\|}$. Thus for all $x_1 \notin \mathscr{B}_2$ we have $\dot{V}_1 < 0$, so the system trajectory converges towards $\mathscr{B}_2$ and enters into this region. Therefore, the ultimate bound is shown to be $\mathscr{B}_2$ as given in (2.22). Hence, the proof is completed. $\qquad\square$

2.3.3 Global Stability of Event-Triggered System

The stability of a ETCS is generally shown using event-triggering rule. If the event-triggering scheme generates a Zeno-free triggering sequence, then the stability of ETCS is assured in some appropriate notion. Following the similar steps as in semi-global triggering scheme, here it is shown the triggering scheme (2.19) ensures global stability.

Theorem 2.4 *Consider the system (1.11). Let $\{t_i\}_{i=0}^{\infty}$ be the sequence of triggering instants generated by the triggering rule (2.19). Then, the inter-event time is always lower bounded by a finite positive quantity for all $x \in \mathbb{R}^n$ and is given as*

$$T_i \geq \frac{1}{\|A\|} \ln\left(1 + \sigma \frac{\|x(t_i)\| + \alpha}{\|c\|(\rho_G(\|x(t_i)\|) + \beta_G)}\right) \tag{2.32}$$

where $\rho_G(\|x(t_i)\|)$ and β_G are defined as

$$\rho_G(\|x(t_i)\|) := \left(\|A - B(c^\top B)^{-1} c^\top A\| + \|B(c^\top B)^{-1} K_2\|\right)\|x(t_i)\| \tag{2.33}$$

and

$$\beta_G := \|B(c^\top B)^{-1} K_1\| + \|B\|d_0. \tag{2.34}$$

Proof We follow the similar steps as in Theorem 2.2. Consider $\Gamma = \{t \in [t_i, t_{i+1}) : \|e(t)\| = 0\}$. For all $t \in [t_i, t_{i+1})\backslash\Gamma$, the error evolution can be obtained by

$$\frac{d}{dt}\|e(t)\| \le \left\|\frac{de(t)}{dt}\right\| = \left\|\frac{dx(t)}{dt}\right\|$$

$$= \left\| Ax(t) - B(c^\top B)^{-1}c^\top Ax(t_i) - \left(B(c^\top B)^{-1}K_1 \right.\right.$$

$$\left.\left. + B(c^\top B)^{-1}K_2\|x(t_i)\|\right)\operatorname{sign}s(t_i) + Bd(t)\right\|. \tag{2.35}$$

Recall $x(t) = x(t_i) - e(t)$, so

$$\frac{d}{dt}\|e(t)\| \le \left\| Ax(t_i) - Ae(t) - B(c^\top B)^{-1}c^\top Ax(t_i) - B(c^\top B)^{-1}K_1\operatorname{sign}s(t_i) \right.$$

$$\left. - B(c^\top B)^{-1}K_2\|x(t_i)\|\operatorname{sign}s(t_i) + Bd(t)\right\|$$

$$\le \|A\|\|e(t)\| + \left\|A - B(c^\top B)^{-1}c^\top A\right\|\|x(t_i)\| + \left\|B(c^\top B)^{-1}K_2\right\|\|x(t_i)\|$$

$$+ \left\|B(c^\top B)^{-1}K_1\right\| + \|B\|d_0$$

$$= \|A\|\|e(t)\| + \rho_G(\|x(t_i)\|) + \beta_G \tag{2.36}$$

where $\rho_G(\|x(t_i)\|)$ and β_G are defined by (2.33) and (2.34), respectively. The solution to the above differential inequality (2.36) can be obtained using comparison Lemma [1] with $\|e(t_i)\| = 0$ as initial condition,

$$\|e(t)\| \le \frac{\rho_G(\|x(t_i)\|) + \beta_G}{\|A\|}\left(e^{\|A\|(t-t_i)} - 1\right) \tag{2.37}$$

for $t \in [t_i, t_{i+1})$. Thus, the time required for $\|e(t)\|$ to grow from 0 to $\sigma(\|x(t_i)\| + \alpha)/\|c\|\|A\|$ is given as

$$\frac{\sigma(\|x(t_i)\| + \alpha)}{\|c\|\|A\|} \le \frac{\rho_G(\|x(t_i)\|) + \beta_G}{\|A\|}\left(e^{\|A\|T_i} - 1\right). \tag{2.38}$$

Rearranging (2.38) gives the relation (2.32). It is seen that the right-hand side of (2.32) is strictly greater than zero for all states in $\mathbb{R}^n$ and hence, Zeno phenomenon is avoided. This implies the event-triggering scheme (2.19) ensures the global stability of the system. This establishes that the event-triggering rule (2.19) makes the system globally stabilizable. Thus, the proof is completed. $\square$

2.3.4 Simulation Results

Consider the same LTI system as given in semi-global stable ETCS

$$\dot{x} = \begin{bmatrix} 0 & 1 \\ 4 & 5 \end{bmatrix} x + \begin{bmatrix} 0 \\ 1 \end{bmatrix}(u + 0.5\sin(10t)).$$

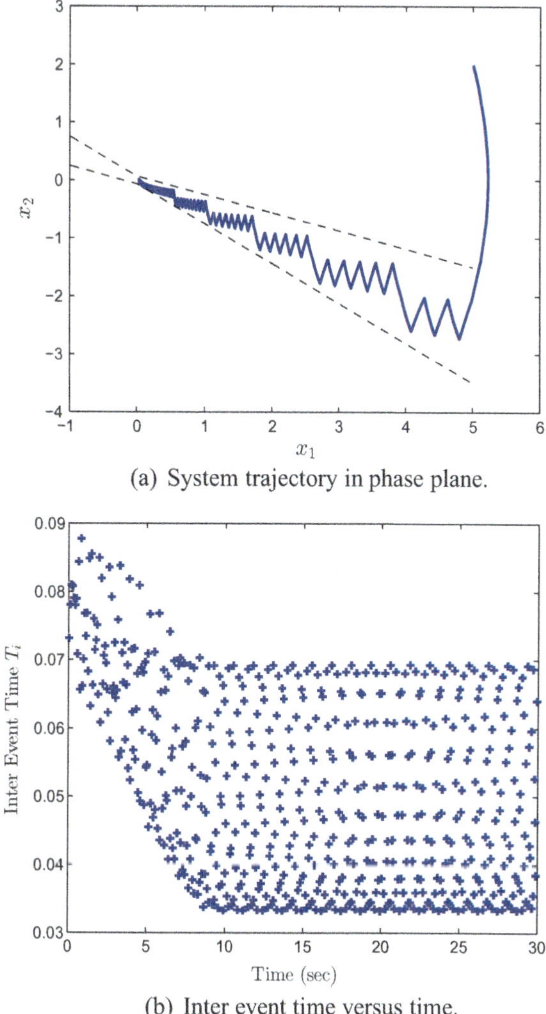

Fig. 2.3 Performance of global event-triggered SMC for LTI systems

(a) System trajectory in phase plane.

(b) Inter event time versus time.

Sliding variable is designed to be $s = \begin{bmatrix} 0.5 & 1 \end{bmatrix} x$, and the corresponding sliding manifold is given as

$$\mathscr{S} = \left\{ x \in \mathbb{R}^2 : s = 0.5x_1 + x_2 = 0 \right\}.$$

The switching control gains are selected as $K_1 = 1.2$ and $K_2 = 1.1$. Event parameters α and σ are designed as 0.5 and 0.85, respectively. The simulation is run in MATLAB with the initial condition $x_0 = \begin{bmatrix} 5 & 2 \end{bmatrix}^\top$.

The performance of the global event-triggered system and control signal is shown in Figs. 2.3 and 2.4, respectively. The state trajectory in state space is attracted towards

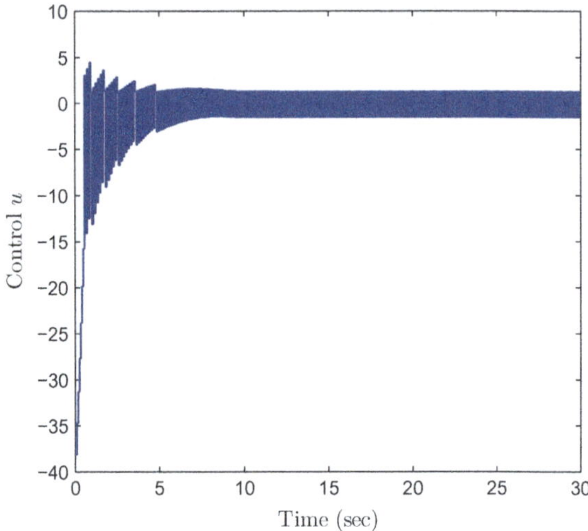

Fig. 2.4 Global event-triggered SMC signal for LTI systems

the sliding manifold and remains bounded once it reaches the vicinity of the sliding manifold as shown in Fig. 2.3a. Since the bound of system trajectory around the manifold is state dependent, the bound decreases as it moves towards the origin with a minimum value at the origin. The variation of sampling interval or inter-event time is plotted in Fig. 2.3b. The positive lower bound for the inter-event time is also seen in Fig. 2.3b as it is proved in Theorem 2.4. However, it is to be noted that inter-event time is not increased much due to large value of switching gain in global triggering scheme. But, this is obtained at the cost of global stability of ETCS. The control signal plot is shown in Fig. 2.4. The control signal is applied to the plant at aperiodic interval, and this value can be increased by setting a large value of α.

2.4 Event-Triggered Sliding Mode Control for Multivariable Systems

Similar to SISO systems, event-triggered SMC can also be designed for multivariable systems, i.e. multi-input multi-output (MIMO) systems. First, SMC for MIMO systems is discussed briefly and then its implementation using event-triggering strategy is presented.

Consider a MIMO system

$$\dot{x} = Ax + B(u + d) \tag{2.39}$$

where $x \in \mathbb{R}^n$ and $u \in \mathbb{R}^m$ are the state and control input, respectively. The matrices A and B are of appropriate dimensions. Here, it is also assumed that the disturbance

$d(t)$ satisfies the matching condition with the control input. Define the multivariable sliding manifold as

$$\mathscr{S} := \left\{ x \in \mathbb{R}^n : s(x) = c^\top x = 0 \right\} \tag{2.40}$$

where $c \in \mathbb{R}^{n \times m}$ and $s = [s_1 \cdots s_m]^\top \in \mathbb{R}^m$. The main difference between SISO manifold, (1.14), and the manifold given by (2.40) is that in the latter case the sliding mode occurs at the intersection of the individual sliding manifold. If $\mathscr{S}_i$ denotes ith sliding manifold for the system (2.39), then the system is said to be in sliding mode if sliding takes place on

$$\mathscr{S} = \bigcap_{i \in \{1,2,\ldots,m\}} \mathscr{S}_i.$$

The control law that brings the sliding mode can be derived as follows. Differentiating the sliding variable, one obtains

$$\dot{s} = c^\top A x + c^\top B u + c^\top B d.$$

The SMC for the system (2.39) is given by

$$u = -(c^\top B)^{-1} \left(c^\top A x + K \operatorname{Sign} s \right) \tag{2.41}$$

where $\operatorname{Sign} s = [\operatorname{sign} s_1 \cdots \operatorname{sign} s_m]^\top$ and $K > \sup_{t \geq 0} \| c^\top B d(t) \|$. Then, it can be easily seen using Lyapunov's direct method that sliding mode occurs in the system in finite-time. Here, in this section, the event-triggered implementation of SMC law (2.41) is discussed. The event-triggered control law can be given as

$$u(t) = -(c^\top B)^{-1} \left(c^\top A x(t_i) + K \operatorname{Sign} s(t_i) \right) \tag{2.42}$$

for all $t \in [t_i, t_{i+1})$. Like SISO, when this control is implemented based on some triggering rule, it results *practical sliding mode* in the system.

2.4.1 Event-Triggered Design of SMC

The following theorem gives the sufficient condition for practical sliding mode of MIMO systems.

Theorem 2.5 *Consider the system (2.39) and the control law (2.42). Let $\alpha > 0$ be given such that*

$$\| c \| \| A \| \| e(t) \| < \alpha \tag{2.43}$$

*holds for all $t \geq 0$. Then, the practical sliding mode is enforced in the system with
the control law* (2.42) *if*

$$K > \sup_{t \geq 0} \left\| c^\top B \right\| \left\| d(t) \right\| + \alpha. \tag{2.44}$$

Proof Consider the Lyapunov function $V(s) = \frac{1}{2} s^\top s$. Taking the derivative of V
with respect to time along the trajectory of (2.39) and using the control law (2.42),

$$\begin{aligned}
\dot{V}(s(t)) &= s^\top(t)\dot{s}(t) \\
&= s^\top(t)\left(c^\top A x(t) + c^\top B u(t_i) + c^\top B d(t)\right) \\
&= s^\top(t)\left(-c^\top A e(t) - K\,\mathrm{Signs}(t_i) + c^\top B d(t)\right).
\end{aligned} \tag{2.45}$$

In the last equation, the relation $e(t) = x(t_i) - x(t)$ is used. Now, we will establish
that sliding manifold is attractive with the control law (2.42). We consider here two
cases: (a) all the sliding trajectories reach the sliding manifolds simultaneously, i.e.
$\mathrm{signs}_j(t_i) = \mathrm{signs}_j(t)$ for all $j = 1, 2, \ldots, m$, and (b) at least one of the sliding
trajectory has crossed the respective sliding manifold, while others do not.

case (a) Since $\mathrm{signs}_j(t_i) = \mathrm{signs}_j(t)$ for all $j = 1, 2, \ldots, m$, all the sliding
trajectory reaches the sliding manifold in finite-time. This implies that $\mathrm{Signs}(t_i) = \mathrm{Signs}(t)$. Using this in the relation (2.45),

$$\begin{aligned}
\dot{V}(s(t)) &= -s^\top(t)c^\top A e(t) - K\|s(t)\|_1 + s^\top(t)c^\top B d(t) \\
&\leq \|s(t)\|\left\|c^\top A e(t)\right\| - K\|s(t)\|_1 + \|s(t)\|\left\|c^\top B d(t)\right\|.
\end{aligned} \tag{2.46}$$

Recall that for any vector $v \in \mathbb{R}^n$, $\|v\|_1 > \|v\|_2$ holds. So, the above relation
reduces to

$$\begin{aligned}
\dot{V}(s) &\leq \|s\|\left\|c^\top A e\right\| - K\|s\| + \|s\|\left\|c^\top B d\right\| \\
&= -\|s\|\left(K - \alpha - \left\|c^\top B d\right\|\right).
\end{aligned} \tag{2.47}$$

Applying (2.44), yields

$$\dot{V}(s) < -\eta\|s\| \tag{2.48}$$

for some $\eta > 0$. This shows that sliding manifold is attractive as long as $\mathrm{Signs}(t_i) = \mathrm{Signs}(t)$. However, when the trajectory reaches the manifold (2.40), the relation
$\mathrm{Signs}(t_i) = \mathrm{Signs}(t)$ does not hold as the control signal is not continuously updated.
So, the trajectory moves away from the manifold after it hits the manifold. Nevertheless, it remains bounded in the vicinity of $\mathscr{S}$ due to relation (2.43). This can be
shown as given in below.

This is the same as finding the maximum deviation of sliding trajectory in one
triggering interval on either sides of $\mathscr{S}$. Using the relation (2.43), it can be shown
that

$$\begin{aligned}
\|s(t) - s(t_i)\| &= \left\| c^\top x(t) - c^\top x(t_i) \right\| \\
&\leq \|c\| \|e(t)\| \\
&< \frac{\alpha}{\|A\|}.
\end{aligned}$$

The maximum bound in the vicinity of sliding manifold can be obtained for the case $s(t_i) = 0$ and is given as

$$\left\{ x \in \mathbb{R}^n : \left\| c^\top x \right\| < \frac{\alpha}{\|A\|} \right\}. \tag{2.49}$$

case (b) Now, we consider the case where $\mathrm{sign} s_j(t_i) \neq \mathrm{sign} s_j(t)$ for some $t \in [t_i, t_{i+1})$ and some $j \in \{1, 2, \ldots, m\}$ in the vicinity of sliding manifold $\mathscr{S}$. It is claimed that even in this case the system trajectory remains bounded within the bound given in (2.49). Let the trajectory $s_j(t)$ has crossed the switching manifold $\mathscr{S}_j$ while all other trajectories $s_k(t)$ have not reached yet the manifolds in the time interval $[t_i, t_{i+1})$ for all $k \in \{1, 2, \ldots, m\}\backslash\{j\}$. To prove this, first we show that all the trajectories which have not reached the respective manifolds move towards manifold. If the sliding trajectories have not reached respective manifolds, then $\mathrm{sign} s_k(t_i) = \mathrm{sign} s_k(t)$ holds for all $k \in \{1, 2, \ldots, m\}\backslash\{j\}$, and hence it implies that $s_k(t)\dot{s}_k(t) < 0$ until it reaches the sliding manifold. Therefore, all the sliding trajectories are attracted towards the sliding manifold even when the sliding trajectory $(s_j(t))$ has already reached the manifold. Now, we establish that the trajectory $s_j(t)$ always remains bounded in the vicinity of sliding manifold as shown below.

The trajectory $s_j(t)$ moves away from the manifold $\mathscr{S}_j$ once it crosses this manifold. But, it remains bounded due to the relation (2.43). To show this,

$$\begin{aligned}
|s_j(t) - s_j(t_i)| &= \left| c_j^\top x(t) - c_j^\top x(t_i) \right| \\
&\leq \|c_j\| \|e(t)\| \\
&< \frac{\alpha}{\|A\|}
\end{aligned}$$

where $c_j^\top$ denotes the jth row of c. This shows that the trajectory will also remain bounded by $\alpha/\|A\|$ in the vicinity of $\mathscr{S}_j$. The same argument can be carried over to all other cases. In a similar way, one can consider the case in which two sliding trajectories have reached the respective manifolds and remains bounded by $\alpha/\|A\|$. Following in the same way, we conclude that the trajectories remain bounded by (2.49). Thus, the proof is completed. $\square$

In multivariable event-triggered SMC, the sliding trajectory remains bounded within a ball (given in (2.49)) around $\mathscr{S}$. The system trajectory moves towards the equilibrium point without leaving the ball due to (2.43). This is similar to the case as seen in SISO system. When the system trajectory is bounded within this ball, the system is said to be in practical sliding mode. The stability of sliding dynamics can be obtained in the same way as in SISO case. The sliding dynamics is given as

$$\dot{x}_1 = \left(A_{11} - A_{12}c_1^\top\right)x_1 + A_{12}s$$
$$x_2 = -c_1^\top x_1 + s.$$

The stability of the above dynamics follows provided the matrix $A_{11} - A_{12}c_1^\top$ is Hurwitz. This is always possible if the pair (A_{11}, A_{12}) is controllable.

2.4.2 Event-Triggering Rule

It is seen from the previous section that the relation (2.43) is essential to ensure practical sliding mode in the system. So, the triggering rule is defined such that this relation holds for all time. Like SISO, the triggering rule here is given as

$$t_{i+1} = \inf\left\{t > t_i : \|c\|\|A\|\|e(t)\| \geq \sigma\alpha\right\} \tag{2.50}$$

for some $\sigma \in (0, 1)$. This generates the triggering instants for updating control signal. Clearly this implies the relation (2.43). So, practical sliding mode would occur in the system with the control law (2.42) and triggering rule (2.50). In the next, we show the stability of ETCS by showing the Zeno-free execution of triggering sequence.

Theorem 2.6 *Consider the system (2.39) and the control law (2.42). Let $\{t_i\}_{i=0}^\infty$ be the triggering sequences generated by (2.43) and $\sigma \in (0, 1)$ be given. Then, the inter-event time, defined by $T_i := t_{i+1} - t_i$, satisfies*

$$T_i \geq \frac{1}{\|A\|}\ln\left(1 + \sigma\frac{\alpha}{\|c\|(\rho_M(\|x(t_i)\|) + \beta_M)}\right) \tag{2.51}$$

where β_M and the real-valued function $\rho_M(\|x(t_i)\|) : \mathbb{R}_{\geq 0} \to \mathbb{R}_{\geq 0}$ are given as

$$\beta_M = \|B(c^\top B)^{-1}\|\sqrt{m}K + \|B\|d_0 \tag{2.52}$$

and

$$\rho_M(\|x(t_i)\|) = \left\|A - B(c^\top B)^{-1}c^\top A\right\|\|x(t_i)\|, \tag{2.53}$$

respectively.

Proof The proof is similar to proof of Theorem 2.2 for SISO case. For the sake of continuity, we present the proof for MIMO case. Define $\Gamma = \{t \in [t_i, t_{i+1}) : \|e(t)\| = 0\}$. Now, differentiating $\|e(t)\|$ for all $t \in [t_i, t_{i+1})\backslash\Gamma$, one obtains

$$\frac{\mathrm{d}}{\mathrm{d}t}\|e(t)\| \leq \left\|\frac{\mathrm{d}}{\mathrm{d}t}e(t)\right\| = \left\|\frac{\mathrm{d}}{\mathrm{d}t}x(t)\right\|$$

$$= \left\|Ax(t) - B(c^\top B)^{-1}c^\top Ax(t_i) - B(c^\top B)^{-1}K\,\mathrm{Signs}\,(t_i) + Bd(t)\right\|$$

$$= \left\|Ax(t_i) - Ae(t) - B(c^\top B)^{-1}c^\top Ax(t_i) - B(c^\top B)^{-1}K\,\mathrm{Signs}\,(t_i) + Bd(t)\right\|$$

$$\leq \|A\|\|e(t)\| + \left\|\left(A - B(c^\top B)^{-1}c^\top A\right)x(t_i)\right\| + \|B(c^\top B)^{-1}\|\sqrt{m}K + \|B\|d_0$$

$$= \|A\|\|e(t)\| + \rho_M(\|x(t_i)\|) + \beta_M \tag{2.54}$$

where β_M and $\rho_M(\|x(t_i)\|)$ are given by (2.52) and (2.53), respectively. The solution to the differential inequality (2.54) can be obtained using comparison Lemma [1] with the initial condition $\|e(t_i)\| = 0$ as

$$\|e(t)\| \leq \frac{\rho_M(\|x(t_i)\|) + \beta_M}{\|A\|}\left(e^{\|A\|(t-t_i)} - 1\right) \tag{2.55}$$

for all time $t \in [t_i, t_{i+1})$. At the triggering instant $\|c\|\|A\|\|e(t_{i+1})\| = \sigma\alpha$, so

$$\frac{\sigma\alpha}{\|c\|\|A\|} = \|e(t_{i+1})\| \leq \frac{\rho_M(\|x(t_i)\|) + \beta_M}{\|A\|}\left(e^{\|A\|T_i} - 1\right). \tag{2.56}$$

The lower bound for T_i follows immediately from the rearrangement of (2.56) is given in (2.51). It is to be noted that this lower bound for T_i is always a finite positive quantity for all finite values of states. In other words, $T_i > a_T$ for some $a_T > 0$ all time $t \in \mathbb{R}_{\geq 0}$ in some domain. Thus, the proof is completed. $\square$

2.4.3 Simulation Results

Consider a numerical example of MIMO system with

$$A = \begin{bmatrix} 1 & 5 & 0 \\ -1 & 2 & 0 \\ 2 & 0 & 1 \end{bmatrix}, \quad B = \begin{bmatrix} 1 & 1 \\ 0 & 1 \\ 1 & 1 \end{bmatrix}, \quad \text{and} \quad C = \begin{bmatrix} 1 & 0 & 1 \\ 0 & 1 & 0 \end{bmatrix}. \tag{2.57}$$

Here, the system has two control inputs and two output variables. Since the two control input appears in the system, the order of the sliding mode dynamics can be obtained as $n - m = 3 - 2 = 1$. First of all, the system is not in regular form and we will transform this system into regular form by the transformation $z = Tx$, where

$$T = \begin{bmatrix} 1 & 0 & -1 \\ 0 & -1 & 1 \\ 0 & 1 & 0 \end{bmatrix}.$$

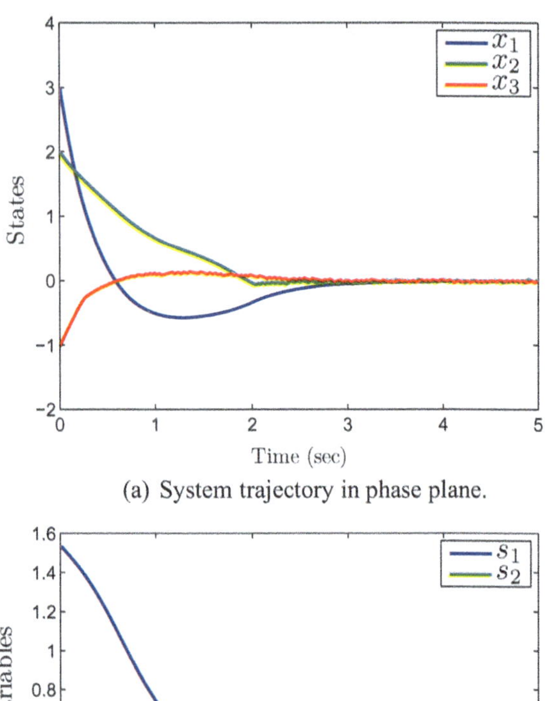

Fig. 2.5 Performance of MIMO event-triggered SMC for LTI systems

(a) System trajectory in phase plane.

(b) Sliding trajectories of the system.

The system (2.57) can be transformed into the system with the transformation matrix as

$$\dot{z}_1 = A_{z_{11}}z_1 + A_{z_{12}}z_2$$
$$\dot{z}_2 = A_{z_{21}}z_1 + A_{z_{22}}z_2 + u$$

where $z = [z_1 \ z_2^\top]^\top$. We proceed to design the event-triggered SMC for this transformed system. The sliding surface matrix is selected as

Fig. 2.6 Performance of MIMO event-triggered SMC for LTI systems

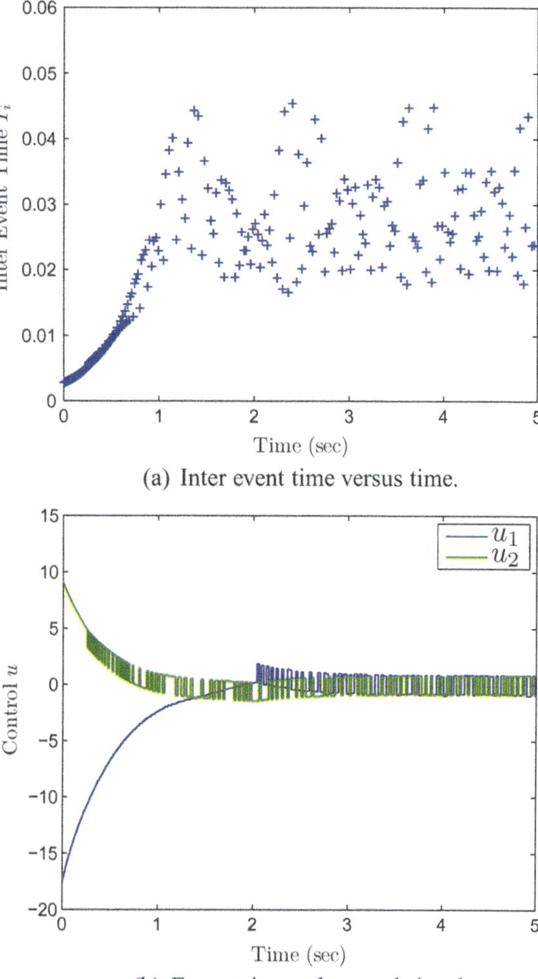

(a) Inter event time versus time.

(b) Event-triggered control signal.

$$c = \begin{bmatrix} -0.1538 & 1 & 0 \\ 0.2308 & 0 & 1 \end{bmatrix}^{\top}$$

such that the reduced system during sliding mode has eigenvalue at -2. Using this, sliding variable is obtained as $s = c^{\top} z$. The event parameters are selected as $\alpha = 0.2$ and $\sigma = 0.9$. The disturbance is taken as $d = \begin{bmatrix} 0.1 + 0.2\cos(5t) & 0.5\sin(10t) \end{bmatrix}^{\top}$ with disturbance bound $d_0 = 0.5831$. Thus, the switching gain K is designed as $K = 0.8 > 0.5831 + 0.2$. Thus, the event-triggered SMC is designed as

$$u(t) = -\left(c^{\top} A_z z(t_i) + K \, \mathrm{Sign} s(t_i) \right)$$

where $A_z = \begin{bmatrix} A_{z11} & A_{z12} \\ A_{z21} & A_{z22} \end{bmatrix}$. The simulation is run in MATLAB with $z(0) = \begin{bmatrix} 3 & 2 & -1 \end{bmatrix}^\top$.

The performance of the event-triggered SMC for the multivariable system is shown in Figs. 2.5 and 2.6. The plots correspond to the transformed system. In Fig. 2.5b, the system trajectories reach the sliding manifold in finite-time and remain bounded within the practical sliding mode band. The state trajectories also converge to origin but ultimately bounded near origin due to switching in control signal as shown in Fig. 2.5a. The plot of inter-event time and the control signals is shown in Fig. 2.6a, b, respectively. It is seen that the inter-event time is increased to a value as high as 0.45, thus reducing frequent control computation. The control becomes discontinuous, when the corresponding sliding trajectories reach the sliding manifold, and remains constants until the triggering instant is generated. This shows that the improved performance of the system is achieved by event-triggering strategy.

2.5 Summary

This chapter mainly deals with the design of event-triggered SMC for linear systems. The SMC is designed first for SISO systems. The stability and triggering rules are explicitly stated for the existence of the practical sliding mode. A variant of triggering rule that ensures the global stability of SISO system is presented. Although global stability is ensured, the performance is slightly deteriorated due to global triggering rule. Then, the design of event-triggered SMC for MIMO system is also discussed. It is shown that similar results are also obtained like SISO.

2.6 Notes and References

The preliminaries on SMC can be found in [2, 3]. The event-triggering-based design of SMC can be found in [4, 5]. The more recent developments in event-triggered SMC can be found in [6–12]. Our approach is based on the preliminary results on event-triggered control found in [13–19]. For different properties of event-triggered system, refer [20].

References

1. H. Khalil, *Nonlinear Systems*, 3rd edn. (Prentice-Hall, Upper Saddle River, 2002)
2. V.I. Utkin, Variable structure systems with sliding modes. IEEE Trans. Autom. Control **22**(2), 212–222 (1977)

3. C. Edwards, S.K. Spurgeon, *Sliding Mode Control: Theory and Applications* (CRC Press, Taylor and Francis Group, 1998)
4. A.K. Behera, B. Bandyopadhyay, Event based robust stabilization of linear systems, in *Proceedings of 40th Annual Conference of the IEEE Industrial Electronics Society, Dallas, USA, Oct–Nov* (2014), pp. 133–138
5. A. Ferrara, G.P. Incremona, L. Magini, Model-based event-triggered robust MPC/ISM, in *Proceedings of 13th European Control Conference, Strausbourg, France* (2014), pp. 2931–2936
6. A.K. Behera, B. Bandyopadhyay, Event based sliding mode control with quantized measurement, in *Proceedings of International Workshop on Recent Advances Sliding Modes, Istanbul, Turkey* (2015), pp. 1–6
7. A.K. Behera, B. Bandyopadhyay, N. Xavier, S. Kamal, Event-triggered sliding mode control for robust stabilization of linear multivarible systems, in *Recent Advances in Sliding Modes: From Control to Intelligent Mechatronics*, ed. by X. Yu, M.Ö. Efe. Studies in Systems, Decision and Control, vol. 24 (Springer International Publishing, Cham, 2015), pp. 155–175
8. A.K. Behera, B. Bandyopadhyay, Self-triggering-based sliding-mode control for linear systems. IET Control Theory Appl. **9**(17), 2541–2547 (2015)
9. A.K. Behera, B. Bandyopadhyay, Event-triggered sliding mode control for a class of nonlinear systems. Int. J. Control **89**(9), 1916–1931 (2016)
10. A.K. Behera, B. Bandyopadhyay, Robust sliding mode control: an event-triggering approach. IEEE Trans. Circuits Syst-II Express Briefs **64**(2), 146–150 (2017)
11. A.K. Behera, B. Bandyopadhyay, Decentralized event-triggered sliding mode control, in *Proceedings of 10th Asian control Conference, Kota Kinabalu, Malaysia* (2015), pp. 968–972
12. A.K. Behera, B. Bandyopadhyay, J. Reger, Discrete event-triggered sliding mode control with fast output sampling feedback, in *Proceedings of 14th IEEE International Workshop on Variable Structure Systems, Nanjing, China* (2016), pp. 148–153
13. K.-E. Årzén, A simple event-based PID controller, in *Proceedings of 14th IFAC World Congrress, Beijing, China* (1999), pp. 423–428
14. K.J. Åström, B. Bernhardsson, Comparison of Riemann and Lebesgue sampling for first order stochastic systems, in *Proceedings of 41st IEEE Conference on Decision and Control, Las Vegas, USA* (2002), pp. 2011–2016
15. P. Tabuada, Event-triggered real-time scheduling of stabilizing control tasks. IEEE Trans. Autom. Control **52**(9), 1680–1685 (2007)
16. W.P.M.H. Heemels, K.H. Johansson, P. Tabuada, An introduction to event-triggered and self-triggered control, in *Proceedings of 51st IEEE Conference of Decision and Control, Hawai, USA* (2010), pp. 3270–3285
17. P. Tabuada, X. Wang, Preliminary results on state-triggered stabilizing control tasks, in *Proceedings of 45th IEEE Conference on Decision and Control, San Deigo, USA* (2006), pp. 282–287
18. J. Lunze, D. Lehmann, A state-feedback approach to event-based control. Automatica **46**(1), 211–215 (2010)
19. W.P.M.H. Heemels, J.H. Sandee, P.P.J.V. Boscho, Analysis of event-driven controllers for linear systems. Int. J. Control **81**(4), 571–590 (2008)
20. D.P. Borger, W.P.M.H. Heemels, Event-separation properties of event-triggered control systems. IEEE Trans. Autom. Control **59**(10), 2644–2656 (2014)

Chapter 3
Event-Triggered Sliding Mode Control for Nonlinear Systems

In this chapter, event-triggered SMC for nonlinear systems is presented. Similar design approaches are used for nonlinear system like LTI systems in the earlier chapter. As a preliminary, first SMC is designed for continuous-time system that yields stable sliding motion. Then, this control is used in designing the event-triggered SMC for nonlinear systems. A sufficient conditions for the stability and existence of practical sliding mode are also given. In practical applications, delays in control implementation are unavoidable that may the affect system performance. The event-triggering strategy is also developed by considering such delays into account.

3.1 System Description

Consider SISO nonlinear systems of the following class as

$$\dot{x}_1 = f_1(x_1) + B_1 x_2 \tag{3.1}$$

$$\dot{x}_2 = f_2(x_1, x_2) + B_2 u + B_2 d \tag{3.2}$$

where $x_1 \in \mathbb{R}^{n-1}$ and $x_2 \in \mathbb{R}$ are the states of the system, u and d are the control input and external disturbance affecting the system, respectively. B_1 and B_2 are matrices of compatible dimensions with $B_2 \neq 0$. The disturbance is assumed to be bounded, i.e. $\sup_{t \geq 0} |d(t)| \leq d_0 < \infty$, and it satisfies the matching condition with respect to the control input. The system considered here is already in standard form to facilitate the control design.

The following assumptions hold for the nonlinear functions $f_1(\cdot)$ and $f_2(\cdot, \cdot)$.

© Springer International Publishing AG, part of Springer Nature 2018

B. Bandyopadhyay and A. K. Behera, *Event-Triggered Sliding Mode Control*, Studies in Systems, Decision and Control 139, https://doi.org/10.1007/978-3-319-74219-9_3

Assumption 3.1 The function $f_1(x_1)$ has an unique equilibrium point and without loss of generality, we assume $f_1(0) = 0$. Further, the system can be represented by both linear and nonlinear terms as $f_1(x_1) = A_1 x_1 + \gamma(x_1)$, where A_1 is the linearized system at the equilibrium point, and the nonlinear component $\gamma(x_1)$ comprises of higher-order terms such that $\frac{\|\gamma(x_1)\|}{\|x_1\|} \to 0$ as $\|x_1\|$ does.

Assumption 3.2 The functions $f_1(\cdot)$ and $f_2(\cdot, \cdot)$ are Lipschitz with respect to its arguments locally in a compact domain $\mathscr{D} \subset \mathbb{R}^n$. So, we write $\|f_1(z_1) - f_1(z_2)\| \le L_1 \|z_1 - z_2\|$ and $|f_2(z_1, y_1) - f_2(z_2, y_2)| \le L_2 \|z_1 - z_2\| + L_2 |y_1 - y_2|$ for some Lipschitz constants L_1 and L_2 of the functions $f_1(\cdot)$ and $f_2(\cdot, \cdot)$, respectively for the corresponding vectors in $\mathscr{D}$.

Assumption 3.3 The pair (A_1, B_1) is controllable.

From the above Assumptions 3.1 and 3.2, it follows that $f(x) = \begin{bmatrix} f_1(x_1) + B_1 x_2 \\ f_2(x_1, x_2) \end{bmatrix}$ is also Lipschitz with Lipschitz constant $L = L_1 + 2L_2 + \|B_1\|$ for any $\mathscr{D} \ni \xi = [z^\top \ y]^\top$ and can be shown as

$$
\begin{aligned}
\|f(\xi_1) - f(\xi_2)\| &= \|f(z_1, y_1) - f(z_2, y_2)\| \\
&\le \|f_1(z_1) - f_1(z_2)\| + \|B_1(y_1 - y_2)\| + |f_2(z_1, y_1) - f_2(z_2, y_2)| \\
&\le L_1 \|z_1 - z_2\| + \|B_1\| |y_1 - y_2| + L_2 \|z_1 - z_2\| + L_2 |y_1 - y_2| \\
&= (L_1 + L_2)\|z_1 - z_2\| + (\|B_1\| + L_2)|y_1 - y_2| \\
&\le (L_1 + L_2)\|\xi_1 - \xi_2\| + (\|B_1\| + L_2)\|\xi_1 - \xi_2\| \\
&= (L_1 + 2L_2 + \|B_1\|)\|\xi_1 - \xi_2\| \\
&= L\|\xi_1 - \xi_2\|.
\end{aligned}
$$

The SMC is designed for this system, and then a sufficient condition is developed for this system when this control is implemented using event-triggered strategy.

3.1.1 Design of Sliding Mode Control

Consider the nonlinear systems given by (3.1)–(3.2). The linear sliding variable is designed for this system and is given by $s = c^\top x$ for some $c \in \mathbb{R}^n$. So, the sliding manifold as

$$
\mathscr{S} \overset{\text{def}}{=} \left\{ x \in \mathbb{R}^n \ : \ s = c^\top x = 0 \right\} \tag{3.3}
$$

where $c = \begin{bmatrix} c_1^\top & 1 \end{bmatrix}^\top$ with $c_1 \in \mathbb{R}^{n-1}$ and $x = \begin{bmatrix} x_1^\top & x_2 \end{bmatrix}^\top$. Differentiating $s = c^\top x$ with respect to time, we obtain

$$\dot{s} = c_1^\top \dot{x}_1 + \dot{x}_2$$
$$= c_1^\top f_1(x_1) + c_1^\top B_1 x_2 + f_2(x_1, x_2) + B_2 u + B_2 d$$
$$= c^\top f(x) + B_2 u + B_2 d. \tag{3.4}$$

The SMC is designed such that it ensures the convergence of system trajectory to sliding manifold (3.3). Then SMC which brings the trajectories of the system to the sliding manifold (3.3) in finite-time can now be designed as

$$u = -B_2^{-1}\left(c^\top f(x) + K \operatorname{sign} s\right) \tag{3.5}$$

where $K > |B_2| d_0$. Then, it can be seen that

$$\dot{s} = -K \operatorname{sign} s + B_2 d.$$

So, the finite-time reachability to the sliding manifold is guaranteed. The main contribution of this chapter is to analyse the event-triggered implementation of the control law (3.5). Since the closed loop system comprises of discontinuous right-hand side, the solutions are understood in the sense of Filippov [1].

In the following, the event-triggered implementation of the control (3.5) is presented. Like linear systems, the triggering rule is developed that guarantees practical sliding mode in the system.

3.2 Event-Triggered Sliding Mode Control

A very few papers have appeared where the discretization of SMC for nonlinear systems is discussed. Similar to the linear systems, in this case also, the discrete implementation of SMC never results exact sliding mode, i.e. $s = 0$. So, the sliding trajectory does not remain on the switching manifold for the discrete case but remain bounded in the vicinity of sliding manifold. In periodic implementation of SMC, the ultimate bound is found to be dependent on sampling interval and disturbance bound. The performance of the system is improved as the sampling interval is decreased. On the other hand, event-triggered implementation, the steady-state bounds are predesigned and thus performance can be improved as desired.

In the following, the event-triggering conditions are developed to bring the system trajectories in the vicinity of sliding manifold in finite-time and maintain the trajectories to remain within this band. Like LTI systems, the trajectory of nonlinear systems (3.1)–(3.2) also remains bounded within some band that is dependent only on some design parameter. Below, a formal definition of sliding mode for nonlinear system (3.1)–(3.2) in the context of event-triggering mechanism is given which we refer as *practical sliding mode*.

Definition 3.1 Let $x(t, x_0)$ be the trajectory of the system (3.1)–(3.2) starting from the initial condition $x_0 = x(t_0, x(t_0))$ and $t > t_0$. The system is said to be in *practical sliding mode* if for given any positive constant ε there exists a finite-time $t_1 \in [t_0, \infty)$ such that the system trajectories reach the sliding manifold in time t_1 and remain bounded by ε in the vicinity of the sliding manifold for all time $t \geq t_1$. The region in the vicinity of sliding manifold where the system trajectories are confined is called *practical sliding band*. The *practical sliding mode* is called *ideal sliding mode* if the trajectories are constrained to $\mathscr{S}$ for all time $t \in [t_1, \infty)$.

Like LTI systems, here also in practical sliding mode the system trajectory remains bounded within a bound which is independent of sampling interval. So, any desired steady-state bound for the sliding motion can be achieved using the event-triggering strategy. Here, the control is held constant until the next triggering instant using ZOH.

Let $\{t_i\}_{i=0}^{\infty}$ be a sequence of triggering instants at which control is updated. Here, $T_i := t_{i+1} - t_i$ denotes the inter-event time. Once the control is updated, it is held constant till the next instant t_{i+1} for all $i \in \mathbb{Z}_{\geq 0}$, i.e. for all $t \in [t_i, t_{i+1})$, the control signal is $u(t) = u(t_i)$. This is known as zero-order hold technique. This is used due to simplest structure in both analysis and design. Define $e(t) := x(t_i) - x(t)$ as the error induced in the system due to discrete implementation of control law such that $e(t_i) = x(t_i) - x(t_i) = 0$. This error e plays a significant role in event-triggering control implementation. The essence of this is that the instant t_{i+1} is determined by observing the evolution of e continuously until it crosses a predefined threshold value. So, this strategy finds more applications where the communication in feedback network is less desired, for example in networked control system. The other advantage of this strategy is that the steady-state bounds can be fixed *a priori* irrespective of the evolution of the state and the disturbance.

Since the control signal is held constant in the time interval between two consecutive triggering instants, i.e. $[t_i, t_{i+1})$, the control signal can be written as

$$u(t) = -B_2^{-1} \left(c^{\top} f(x(t_i)) + K \operatorname{signs}(t_i) \right) \tag{3.6}$$

for all $t \in [t_i, t_{i+1})$. The objective here is to design the switching gain such that the system trajectories remain bounded in finite-time. This is obtained in a similar manner as it is done for SISO.

Theorem 3.1 *Consider the system (3.1)–(3.2) and the control law (3.6). Let $\alpha > 0$ be given such that*

$$L \, \|c\| \, \|e(t)\| < \alpha \tag{3.7}$$

for all time $t \geq 0$. Then, the practical sliding mode occurs in the system if the gain K is selected as

$$K > |B_2| \, d_0 + \alpha. \tag{3.8}$$

Proof To prove, we consider the Lyapunov function $V = \frac{1}{2}s^2$. Differentiating V with respect to time $t \in [t_i, t_{i+1})$ along the system trajectories of (3.1)–(3.2), we obtain

$$\dot{V}(s) = s\dot{s}$$
$$= s\left(c^\top f(x) + B_2 u + B_2 d\right). \tag{3.9}$$

Using (3.6) in the above relation yields

$$\dot{V}(s(t)) = s(t)\left(c^\top f(x(t)) - c^\top f(x(t_i)) - K\operatorname{sign}s(t_i) + B_2 d(t)\right)$$
$$\leq -s(t)K\operatorname{sign}s(t_i) + |s(t)|\,\left\|c^\top f(x(t)) - c^\top f(x(t_i))\right\| + |s(t)|\,|B_2|\,d_0$$
$$\leq -s(t)K\operatorname{sign}s(t_i) + |s(t)|\,\|c\|\,\|f(x(t)) - f(x(t_i))\| + |s(t)|\,|B_2|\,d_0$$
$$\leq -s(t)K\operatorname{sign}s(t_i) + |s(t)|L\,\|c\|\,\|x(t) - x(t_i)\| + |s(t)|\,|B_2|\,d_0. \tag{3.10}$$

It is to be noted that the sign of sliding variable does not change until the trajectory reaches the sliding manifold. So, it implies that $\operatorname{sign}s(t_i) = \operatorname{sign}s(t)$. With this, the first term in (3.10) can be written as $-K|s(t)|$. Using this along with the relations (3.7) and (3.8) in (3.10), we obtain

$$\dot{V}(s(t)) \leq -|s(t)|K + |s(t)|\alpha + |s(t)|\,|B_2|\,d_0$$
$$= -|s(t)|\,(K - \alpha - |B_2|\,d_0)$$
$$< -\eta|s(t)| \tag{3.11}$$

for some $\eta > 0$. This shows that the trajectory moves towards the sliding manifold during the time interval $[t_i, t_{i+1})$ for some $i \in \mathbb{Z}_{\geq 0}$. This process continues for the subsequent triggering intervals as long as $\operatorname{sign}s(t_i) = \operatorname{sign}s(t)$. Eventually, the trajectory hits the manifold in finite-time due to (3.11). However, it is not guaranteed that the trajectory remains on this manifold as the control signal is not applied continuously. So, the trajectory crosses the manifold after hitting it. But, it does not go unbounded since the relation (3.7) holds which is shown below.

We obtain the maximum deviation of sliding trajectory in any time interval $[t_i, t_{i+1})$. This can be written using (3.7) as

$$|s(t_i) - s(t)| = \left|c^\top x(t_i) - c^\top x(t)\right|$$
$$\leq \|c\|\,\|e(t)\|$$
$$< \frac{\alpha}{L}.$$

The maximum value of practical sliding mode band can be obtained if the triggering takes place when the trajectory just reaches $\mathscr{S}$, i.e. $s(t_i) = 0$, and thus, the band is given as

$$\Omega = \left\{ x \in \mathscr{D} : |s| = \left|c^{\top} x\right| < \frac{\alpha}{L} \right\}. \tag{3.12}$$

This shows that the system trajectory is ultimately bounded in the region Ω. Thus, the proof is completed. $\qquad\square$

The above result shares some similarity with linear systems. First is the relation (3.7) which is essential for the existence of practical sliding mode in the nonlinear systems. The Lipschitz constant appearing here corresponds to the induced norm of the system matrix of LTI systems. Another similarity is that the practical sliding mode band obtained from the similar relation (3.7) as it is obtained for LTI systems.

3.2.1 Stability of Sliding Motion

The stability of the system dynamics is governed by the sliding variable parameter, i.e. c. The sliding motion dynamics during the practical sliding mode can be given as

$$\dot{x}_1 = \left(A_1 - B_1 c_1^{\top}\right) x_1 + \gamma(x_1) + B_1 s \tag{3.13}$$

$$x_2 = -c_1^{\top} x_1 + s. \tag{3.14}$$

The vector c_1 is always designed such that the matrix $A_c = A_1 - B_1 c_1^{\top}$ is Hurwitz. This is possible due to Assumption 3.3. From the above dynamics, it can be seen that if x_1 is bounded, x_2 also remains bounded due to (3.14). In the following, the stability of the sliding dynamics is given.

Proposition 3.1 *Consider the system* (3.13) *and* (3.14). *Then, the system trajectories are ultimately bounded if the matrix A_c is Hurwitz and for some $\ell > 0$, $\gamma(\|x_1\|) \le \ell\|x_1\|$ for all $x \in \mathscr{D}$.*

Proof Consider the Lyapunov function $V_1 = x_1^{\top} P x_1$ where $P > 0$. Differentiating V_1 with respect to time along the system trajectories of (3.13), one can obtain

$$\begin{aligned}
\dot{V}_1 &= \dot{x}_1^{\top} P x_1 + x_1^{\top} P \dot{x}_1 \\
&= x_1^{\top} (A_c^{\top} P + P A_c) x_1 + 2 x_1^{\top} P \gamma(x_1) + 2 x_1^{\top} P B_1 s.
\end{aligned} \tag{3.15}$$

Since the matrix A_c is Hurwitz, $A_c^{\top} P + P A_c + Q = 0$ for any given $Q > 0$. Then, it implies

$$\dot{V}_1 = -x_1^{\top} Q x_1 + 2 x_1^{\top} P \gamma(x_1) + 2 x_1^{\top} P B_1 s. \tag{3.16}$$

This can be further reduced to

$$\dot{V}_1 \le -\lambda_{\min}\{Q\}\|x_1\|^2 + 2\|x_1\|\|P\|\|\gamma(x_1)\| + 2\|x_1\|\|P B_1\||s|. \tag{3.17}$$

From the fact of Assumption 3.1, for any $\ell_1 > 0$ there exists $r_1 > 0$ such that $\|\gamma(x_1)\| \le \ell_1 \|x_1\|$ for $\|x_1\| \le r_1$. Define $r^\star := \sup_{x \in \mathscr{D}} \|x_1\|$. Then, $\ell > 0$ can be found such that $\|\gamma(x_1)\| \le \ell \|x_1\|$ for $\|x_1\| \le r$ with $r \ge r^\star$. This is possible by selecting an appropriate domain $\mathscr{D}$. Now, we can write

$$\begin{aligned}
\dot{V}_1 &\le -\lambda_{\min}\{Q\}\|x_1\|^2 + 2\ell\|P\|\|x_1\|^2 + 2\|PB_1\|\|x_1\|\,|s| \\
&= -\left(\lambda_{\min}\{Q\} - 2\ell\|P\|\right)\|x_1\|^2 + 2\|PB_1\|\|x_1\|\,|s| \\
&= -\|x_1\|\left(\lambda_{\min}\{Q\} - 2\ell\|P\|\right)\left(\|x_1\| - \Theta|s|\right).
\end{aligned} \tag{3.18}$$

where $\Theta := \frac{2\|PB_1\|}{\lambda_{\min}\{Q\}-2\ell\|P\|}$. Choosing $\ell < \lambda_{\min}\{Q\}/2\|P\|$ in the domain $\mathscr{D}$, it can be seen that $\dot{V}_1 < 0$ for all x_1 outside the ball $\{x_1 \in \mathscr{D}_1 : \|x_1\| \le \Theta|s|\}$ where $\mathscr{D}_1 \subset \mathscr{D}$. Thus, the closed loop system (3.13) is ISS with respect to s. The ultimate bound of the system trajectories of (3.13) can be obtained from (3.12) as

$$\left\{x_1 \in \mathscr{D}_1 : \|x_1\| < \Theta\frac{\alpha}{L}\right\}.$$

This completes the proof. $\qquad\square$

3.3 Event-Triggering Strategy

The triggering condition must ensure the stability of the system. From the previous section, it is seen that the relation (3.7) is a sufficient condition for the existence of practical sliding mode. So, this relation must be satisfied for all time to guarantee the system stability. In other words, the triggering scheme should be developed such that this relation always holds. Thus, the triggering scheme is formulated as

$$t_{i+1} = \inf\left\{t > t_i : L\|c\|\|e(t)\| \ge \sigma\alpha\right\} \tag{3.19}$$

where $\sigma \in (0, 1)$. This triggering scheme ensures the relation

$$L\|c\|\|e(t)\| < \sigma\alpha \tag{3.20}$$

always true for all time $t \ge 0$.

Let $\{t_i\}_{i=0}^{\infty}$ be a sequence of triggering instants. For stability of the ETCS, there must exists a positive lower bound between any two consecutive triggering instants, i.e. Zeno execution of the sequence is avoided. Indeed, this is essential for digital processors to execute the control tasks. Otherwise, the system may become unstable.

In practical situations, the control law is applied to the plant at the discrete-time instants given by the triggering sequence $\{t_i\}_{i=0}^{\infty}$ provided there is no delay associated with the control execution. Sometimes the delay is discarded if it is very small to affect the system performance. With this assumption on delay-free control execution, we show that the triggering sequence generated by (3.19) is a Zeno-free sequence in the following Theorem. Before going to prove this, first we write the system dynamics (3.1)–(3.2) as given in the following form

$$\dot{x} = f(x) + Bu + Bd$$

where $B = \begin{bmatrix} \mathbf{0}^\top & B_2 \end{bmatrix}^\top$ where $\mathbf{0}$ being a column vector of dimension $n - 1$ with all entries are zero.

Theorem 3.2 *Consider the system (3.1)–(3.2) and the control law (3.6). Let $\{t_i\}_{i=0}^{\infty}$ be a triggering sequence satisfies generated by the triggering rule given by (3.19). Then, the inter-execution time T_i is lower bounded by a positive value and is given as*

$$T_i \geq \frac{1}{L} \ln\left(1 + \sigma \frac{\alpha}{\|c\| \left(\rho_N(\|x(t_i)\|) + \beta_N\right)}\right) \tag{3.21}$$

where κ is defined as

$$\beta_N = \left\| B B_2^{-1} \right\| K + |B_2| d_0 \tag{3.22}$$

and the real-valued function $\rho_N(\|x(t_i)\|) : \mathbb{R}_{\geq 0} \mapsto \mathbb{R}_{\geq 0}$ is a class-$\mathscr{K}_\infty$ and is given as

$$\rho_N(\|x(t_i)\|) = L \left(1 + \left\| B B_2^{-1} c^\top \right\|\right) \|x(t_i)\|. \tag{3.23}$$

Proof In the proof, we first find the time required by $\|e\|$ to grow from zero to $\sigma\alpha/\|c\|L$ and then show that it is lower bounded from zero. Define $\Gamma := \{t \in [t_i, t_{i+1}) : \|e(t)\| = 0\}$. Then, T_i can be derived to be the time taken by $\|e\|$ to grow from 0 to $\sigma\alpha/L\|c\|$. For $t \in [t_i, t_{i+1})\backslash\Gamma$

$$\begin{aligned}
\frac{\mathrm{d}}{\mathrm{d}t}\|e(t)\| &\leq \left\|\frac{\mathrm{d}}{\mathrm{d}t}e(t)\right\| = \left\|\frac{\mathrm{d}}{\mathrm{d}t}x(t)\right\| \\
&= \|f(x(t)) + Bu(t) + Bd(t)\| \\
&= \left\| f(x(t)) - B B_2^{-1} c^\top f(x(t_i)) - B B_2^{-1} K \operatorname{sign} s(t_i) + Bd(t)\right\|. \tag{3.24}
\end{aligned}$$

That last equality is obtained by substituting the control expression (3.6). Using $x(t) = x(t_i) - e(t)$, we can reduce (3.24) further as

$$\frac{\mathrm{d}}{\mathrm{d}t}\|e(t)\| \le L\|x(t)\| + \left\|BB_2(c^\top B)^{-1}c^\top f(x(t_i))\right\| + \left\|BB_2^{-1}\right\| K + \|B_2 d(t)\|$$

$$\le L(\|x(t_i)\| + \|e(t)\|) + L\left\|BB_2(c^\top B)^{-1}c^\top\right\|\|x(t_i)\| + \left\|BB_2^{-1}\right\| K + |B_2|d_0$$

$$= L\|e(t)\| + \left(1 + \left\|BB_2^{-1}c^\top\right\|\right) L\|x(t_i)\| + \kappa$$

$$= L\|e(t)\| + \rho_N(\|x(t_i)\|) + \beta_N \tag{3.25}$$

where β_N and $\rho_N(\|x(t_i)\|)$ are defined as in (3.22) and (3.23), respectively. For $t \in [t_i, t_{i+1})\backslash \Gamma$, the solution to the above differential inequality is obtained by invoking the comparison Lemma [2] with the initial condition $\|e(t_i)\| = 0$ as

$$\|e(t)\| \le \frac{\rho_N(\|x(t_i)\|) + \beta_N}{L}\left(e^{L(t-t_i)} - 1\right). \tag{3.26}$$

The triggering instant t_{i+1} is triggered as soon as (3.19) is satisfied. So, we write (3.26) as

$$\frac{\sigma\alpha}{L\|c\|} = \|e(t_{i+1})\| \le \frac{\rho_N(\|x(t_i)\|) + \beta_N}{L}\left(e^{LT_i} - 1\right). \tag{3.27}$$

Rearranging (3.27) gives the expression (3.21) for inter-execution time. Now, it remains to show that it is lower bounded by some finite-positive quantity. Note that $\rho(\|x(t_i)\|)$ and κ both are finite and positive quantities. Thus, it implies that T_i is always bounded below by a positive finite quantity. This ends the proof. $\qquad\square$

Remark 3.1 The result in Theorem 3.2 ensures $T_i > 0$ in the domain $\mathscr{D}$. So, the triggering scheme proposed for the class of nonlinear system is a local one. However, the results can be valid in a larger domain if the system is stabilizable by SMC in that domain in continuous-time setting. But, the global stability cannot be achieved as the triggering mechanism does not have a global property.

3.3.1 Design of Event-Triggering Scheme with Constraints

In the above, it is discussed that the triggering sequences are always separated from its immediate previous instant by a finite-positive quantity. In the following, we analyse how to ensure the inter-execution time is greater than a given positive bound for T_i. This is important because in practice, the digital processors can only entertain the control task subject to some minimum time period corresponding to the processor bandwidth. So, if this can be ensured, Zeno execution can be avoided in more practical scenario. It is well known that the choice of α determines the steady-state bound of the system, so it must be chosen sufficiently large such that no accumulation of control executions occurs, i.e. triggering instants generated are larger than minimum given time period. For instance, for the given small value of α, the next triggering instant can be below the sampling intervals corresponding to bandwidth of the processor. If such a situation arises, the control will not be executed until the triggering instants go beyond the minimum time period of processor speed, and eventually may result in instability of the system. In other words, T_i must be large enough to ensure such problem is avoided. We provide below the condition on α to ensure the inter-execution time T_i is larger than the processor bandwidth for all $i \in \mathbb{Z}_{\geq 0}$. The following Theorem gives a lower bound on α to ensure the constraint of the processor is respected.

Theorem 3.3 *Consider the system (3.1)–(3.2) and the control law (3.6). Let $\sigma \in (0, 1)$ be given and T_o as the minimum time period of the processor for the control task to be executed. If*

$$\alpha \geq \alpha_{\min} \tag{3.28}$$

then T_i is always greater than T_o where

$$\alpha_{\min} = \frac{(\rho_N(\|x(t_i)\|) + \beta_N)\,\|c\|}{\sigma}\left(e^{LT_o} - 1\right). \tag{3.29}$$

Proof The idea of this proof is to estimate the minimum increment time of $\|e(t)\|$ in the interval $[t_i, t_{i+1})$ and then to show this minimum time is always smaller than or equal to $T_i = t_{i+1} - t_i$. First few steps follow the proof of Theorem 3.2. Choose $\alpha \geq \alpha_{\min}$. Then the evolution of $\|e\|$ for $t \in [t_i, t_i + T_o)\backslash\Gamma$ can be given from (3.26) as

$$\|e(t)\| \leq \frac{\rho_N(\|x(t_i)\|) + \beta_N}{L}\left(e^{L(t-t_i)} - 1\right). \tag{3.30}$$

Let the error e grows from zero to $\sigma\alpha/L\|c\|$ during the time interval $[t_i, t_{i+1})$. Then from (3.21), the inter-execution time can be written as

$$T_i \geq \frac{1}{L}\ln\left(1 + \sigma\frac{\alpha}{\|c\|\,(\rho_N(\|x(t_i)\|) + \beta_N)}\right). \tag{3.31}$$

Since the right-hand side of (3.31) is a monotonic function with α, so with $\alpha \geq \alpha_{\min}$, it yields

$$T_i \geq \frac{1}{L}\ln\left(1 + \sigma\frac{\alpha_{\min}}{\|c\|\,(\rho_N(\|x(t_i)\|) + \beta_N)}\right)$$
$$= T_o. \tag{3.32}$$

This is true for all control tasks as $\|x(t_i)\|$ decreases or at least remains bounded. So, the inter-execution time is larger than the minimum time period corresponding to processor bandwidth for all time. Hence, the proof is completed. $\qquad\square$

The next part discusses how to design α such that the minimum given lower bound of inter-execution time is guaranteed. The design methodology for the value of α

Algorithm 1 Design of the value of α

1: Start
2: Initialize σ, $T_o^\star$ and $\alpha_{\min}^\star$
3: Select $\alpha^\star \geq \alpha_{\min}^\star$ and calculate $K^\star$ from (3.8) and then β_N from (3.22)
4: Compute T_o from (3.29)
5: **if** $T_o > T_o^\star$ **then**
6:     Set $\alpha_{\min} \leftarrow \alpha_{\min}^\star$
7:     Set $\alpha \leftarrow \alpha^\star$ and $K \leftarrow K^\star$
8: **else**
9:     **go to** 2:
10: **end**
11: **return** α and K
12: End

---

can be obtained from (3.28). To calculate this, we need the value of $\alpha_{\min}$. It can be easily seen that $\alpha_{\min}$ depends on $\rho_N(\|x(t_i)\|)$, β_N and T_o. However, κ in turn depends on α through K. Thus, in order to get the suitable value of α, a systematic design procedure must be given that guarantees no accumulation of inter-execution times. The following algorithm provides a procedure to select the value of α.

Remark 3.2 The initial guess of $\alpha_{\min}^\star$ is the crucial part of the algorithm, and it must be chosen sufficiently large to generate T_i larger than minimum inter-execution time T_o of processor. However, the higher value of $\alpha_{\min}$ might increase the steady-state bound of the sliding trajectory. On the other hand, too small value may generate *Zeno phenomenon* of inter-execution times. So, a trade-off for optimal value of α is chosen for optimum performance of system subject to desired steady-state bound. The Algorithm 1 may be used for different set of values of α, and an optimum value may be selected based for different steady-state values.

3.4 Event-Triggering with Delay

Here, event-triggering scheme by taking the delay into account is investigated. In the previous section, all the analyses are based on the fact that the control is applied to the plant at the instant it is updated. However, in practical applications, it is very unrealistic situation. Given the fact that the delay associated with the control task is very small, we can neglect the effect of delay and the previous analysis can be borrowed to design the event-triggered SMC. On the other hand, if the delay is significantly large compared to the sampling intervals of event-triggering scheme, we cannot avoid the unaccounted delay in the analysis. So, in this section, we deal with the design of event-triggering-based SMC by considering the delay into account.

Let δ_i be the delay associated with the triggering instant t_i. Then, $\{t_i + \delta_i\}_{i=0}^\infty$ denotes the sequence of time instants for updating control signal corresponding to the triggering sequence $\{t_i\}_{i=0}^\infty$. Then, the control signal is applied to the plant at the time instant $t_i + \delta_i$ corresponding to the triggering instant t_i as shown in Fig. 3.1. Due

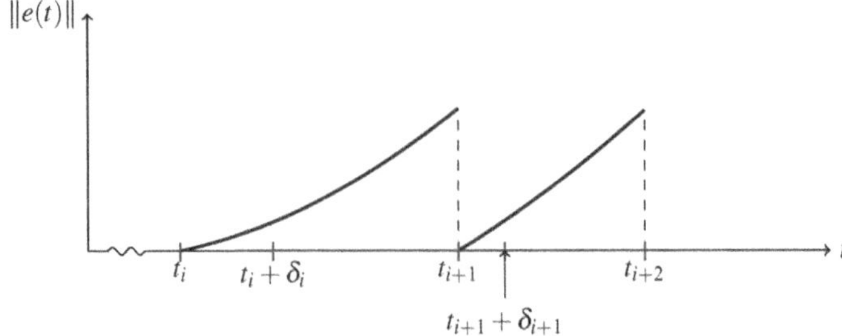

Fig. 3.1 Evolution of $\|e(t)\|$ with time at event-triggering block

to this delay the control is not updated at the sampling instants, so the error of the system grows till the next control signal is applied to the plant. We call the triggering sequence $\{t_{i+1}\}_{i=0}^{\infty}$ is *admissible* if $t_{i+1} > t_i + \delta_i$ for all $i \in \mathbb{Z}_{\geq 0}$. Figure 3.1 shows the situation where the generated triggering instants are admissible. We discuss in the following, under what condition the triggering sequence is admissible and the triggering rule guarantees system stability. In the next, we give the upper bound on the delay such that the closed loop system is stable and practical sliding mode is enforced. Also, the design of triggering scheme is discussed so that the triggering rule (3.19) generates an admissible triggering sequence.

During the time interval $[t_i, t_i+\delta_i)$, the control signal $u(t_i)$ is not applied. However, the previous control signal $u(t_{i-1})$ is still acting on the system during this time interval. So, for the delayed control signal in the time interval $[t_i, t_{i+1} + \delta_{i+1})$, the control law can be given as

$$u(t) = -B_2^{-1}\left(c^{\top} f(x(t_{i-1})) + K \operatorname{signs}(t_{i-1})\right) \tag{3.33}$$

$$u(t) = -B_2^{-1}\left(c^{\top} f(x(t_i)) + K \operatorname{signs}(t_i)\right) \tag{3.34}$$

for all $t \in [t_i, t_i + \delta_i)$ and $t \in [t_i + \delta_i, t_{i+1} + \delta_{i+1})$, respectively. It is shown that during $[t_i, t_i + \delta_i)$, the control is not updated with the current sampled state $x(t_i)$, whereas the updated control is continuously applied for all $t \in [t_i + \delta_i, t_{i+1} + \delta_{i+1})$. The sufficient condition for admissible triggering sequence is given in the following Theorem.

Theorem 3.4 *Consider the system* (3.1)–(3.2) *and the control law* (3.33)–(3.34). *Let* $\sigma_1 \in (0, 1)$ *and* $\sigma \in (\sigma_1, 1)$ *be given. Define*

$$\varepsilon_i := \frac{1}{L} \ln\left(1 + \sigma_1 \frac{\alpha}{\|c\|(\rho_{N_1}(\|x(t_i)\|, \|x_{i-1}\|) + \beta_N)}\right) \tag{3.35}$$

and

$$\vartheta_i := \frac{1}{L} \ln\left(1 + \frac{\sigma\alpha - \sigma_1\alpha}{\|c\| (\rho_N(\|x(t_i)\|) + \beta_N) + \sigma_1\alpha}\right) \tag{3.36}$$

where $\rho_{N_1}(\|x(t_i)\|, \|x(t_{i-1})\|)$ is defined as

$$\rho_{N_1}(\|x(t_i)\|, \|x(t_{i-1})\|) := L\left(\|x(t_i)\| + \|BB_2^{-1}c^\top\| \|x(t_{i-1})\|\right), \qquad (3.37)$$

and $\rho_N(\|x(t_i)\|)$ and β_N are given by (3.23) and (3.22), respectively. Then, the triggering sequence is admissible if

$$\delta_i \le \varepsilon_i \qquad (3.38)$$

for all $i \in \mathbb{Z}_{\ge 0}$.

Proof The proof follows in two steps, (i) for $t \in [t_i, t_i + \delta_i)$ and (ii) for $t \in [t_i + \delta_i, t_{i+1} + \delta_{i+1})$. Define $\Gamma_1 = \{t \in [t_i, t_i + \delta_i) : \|e(t)\| = 0\}$. Then, for all $t \in [t_i, t_i + \delta_i)\backslash\Gamma_1$, we find the evolution of $\|e\|$, thus

$$\begin{aligned}
\frac{\mathrm{d}}{\mathrm{d}t}\|e(t)\| &\le \left\|\frac{\mathrm{d}}{\mathrm{d}t}e(t)\right\| = \left\|\frac{\mathrm{d}}{\mathrm{d}t}x(t)\right\| \\
&= \|f(x(t)) + Bu(t) + Bd(t)\| \\
&= \left\|f(x(t)) - BB_2^{-1}c^\top f(x(t_{i-1})) - BB_2^{-1}K\,\mathrm{signs}(t_{i-1}) + Bd(t)\right\|.
\end{aligned}$$

$$(3.39)$$

Now using $x(t) = x(t_i) - e(t)$ in the above relation and simplifying it further, one can obtain

$$\begin{aligned}
\frac{\mathrm{d}}{\mathrm{d}t}\|e(t)\| &\le L\|x(t)\| + \|BB_2^{-1}c^\top\| L\|x(t_{i-1})\| + \|BB_2^{-1}\| K + \|B_2 d(t)\| \\
&\le L(\|x(t_i)\| + \|e(t)\|) + \|BB_2^{-1}c^\top\| L\|x(t_{i-1})\| + \|BB_2^{-1}\| K + |R_2|d_0 \\
&= L\|e(t)\| + L\left(\|x(t_i)\| + \|BB_2^{-1}c^\top\| \|x(t_{i-1})\|\right) + \beta_N \\
&= L\|e(t)\| + \rho_{N_1}(\|x(t_i)\|, \|x(t_{i-1})\|) + \beta_N
\end{aligned} \qquad (3.40)$$

where $\rho_{N_1}(\|x(t_i)\|, \|x(t_{i-1})\|)$ and β_N are given by (3.37) and (3.22), respectively. The solution to the differential inequality (3.40) can be obtained using comparison Lemma [2] with $\|e(t_i)\| = 0$ as

$$\|e(t)\| \le \frac{\rho_{N_1}(\|x(t_i)\|, \|x(t_{i-1})\|) + \beta_N}{L}\left(e^{L(t-t_i)} - 1\right) \qquad (3.41)$$

for all $t \in [t_i, t_i + \delta_i)$. Since the right-hand side of (3.41) is a monotonic function with t, using (3.38) and (3.35) the above relation can be deduced to

$$\|e(t)\| \leq \frac{\rho_{N_1}(\|x(t_i)\|, \|x(t_{i-1})\|) + \beta_N}{L} \left(e^{L(t-t_i)} - 1\right)$$

$$< \frac{\rho_{N_1}(\|x(t_i)\|, \|x(t_{i-1})\|) + \beta_N}{L} \left(e^{L\delta_i} - 1\right)$$

$$\leq \frac{\rho_{N_1}(\|x(t_i)\|, \|x(t_{i-1})\|) + \beta_N}{L} \left(e^{L\varepsilon_i} - 1\right)$$

$$= \frac{\sigma_1 \alpha}{L\|c\|}$$

for all $t \in [t_i, t_i + \delta_i)$. This gives the maximum value of error during the delay δ_i. That means, for all $t \in [t_i, t_i + \delta_i)$ the error can grow at most $\sigma_1 \alpha / L\|c\|$ provided (3.38) is true.

To proceed for the proof of (ii), we refer the relation (3.25) from Theorem 3.2. The solution to (3.25) with $\|e(t_i + \delta_i)\| = \frac{\sigma_1 \alpha}{L\|c\|}$ as initial condition can be shown as

$$\|e(t)\| \leq \frac{\rho_N(\|x(t_i)\|) + \beta_N}{L} \left(e^{L(t-t_i-\delta_i)} - 1\right) + \frac{\sigma_1 \alpha}{L\|c\|} e^{L(t-t_i-\delta_i)} \tag{3.42}$$

for all $t \in [t_i+\delta_i, t_{i+1}+\delta_{i+1})$. However, the triggering instant is generated whenever $\|e(t)\| = \frac{\sigma\alpha}{L\|c\|}$ according to (3.19). So, at the moment triggering occurs, it can be written as

$$\frac{\sigma\alpha}{L\|c\|} = \|e(t_{i+1})\| \leq \frac{\rho_N(\|x(t_i)\|) + \beta_N}{L} \left(e^{L(t_{i+1}-t_i-\delta_i)} - 1\right) + \frac{\sigma_1 \alpha}{L\|c\|} e^{L(t_{i+1}-t_i-\delta_i)}.$$

Rearranging the above relation to obtain the expression for $t_{i+1} - t_i - \delta_i$, we arrive at the following

$$t_{i+1} - t_i - \delta_i \geq \frac{1}{L} \ln\left(1 + \frac{\sigma\alpha - \sigma_1\alpha}{\|c\|\,(\rho_N(\|x(t_i)\|) + \beta_N) + \sigma_1\alpha}\right) =: \vartheta_i.$$

Now recall that $\sigma > \sigma_1$, so we conclude from (3.36) that $\vartheta_i > 0$ for all $i \in \mathbb{Z}_{\geq 0}$. This implies that the triggering rule (3.19) generates an admissible triggering sequence, i.e. $t_{i+1} > t_i + \delta_i$ for all $i \in \mathbb{Z}_{\geq 0}$. Thus, for all delays satisfying (3.38) and $\sigma > \sigma_1$, we ensure that the triggering sequences are admissible with $\|e(t)\| \leq \sigma\alpha/L\|c\|$ for all $t \in [t_i, t_{i+1} + \delta_{i+1})$ and $i \in \mathbb{Z}_{\geq 0}$. This ends the proof. $\square$

In the next, we discuss the effect of delay in the control implementation on the size of sliding band. Note that Theorem 3.4 guarantees that the triggering rule is always satisfied since the relation (3.20) holds for all time. So, as long as the condition in Theorem 3.4 is satisfied, the stability of the closed loop system under the proposed event-triggering scheme also holds. However, due to this delayed control update, the size of sliding mode band is affected; larger the delay implies high magnitude of the practical sliding mode band. This happens because even if the states are sampled at t_i the control is updated at $t_i + \delta_i$. Since there is no update in control signal in the interval $[t_i + \delta_i, t_{i+1} + \delta_{i+1})$, the error in the plant grows and becomes maximum during the interval. Denote plant error by $e_p(t) = x(t_i) - x(t)$ for time $t \in [t_i + \delta_i, t_{i+1} + \delta_{i+1})$. During the time interval $[t_i, t_i + \delta_i)$, the error $\|e\|$ grows from zero to $\frac{\sigma_1 \alpha}{L\|c\|}$ under the conditions of Theorem 3.4. However, $\|e_p\|$ increases to $\frac{\sigma\alpha}{L\|c\|}$ during this time interval and eventually goes beyond this value until the control signal is applied to the plant. In spite of this, it can be shown that the error in the plant remains bounded for all time.

Proposition 3.2 *Consider the system (3.1)–(3.2) and the control law (3.33)–(3.34). Let $\sigma_1 \in (0, 1)$ and $\sigma \in (\sigma_1, 1)$ be given. Assume that (3.7) holds for all time and the closed loop system is stable. Then, the error in the plant remains bounded by $(\sigma + \sigma_1)\frac{\alpha}{L\|c\|}$, i.e. $\|e_p(t)\| < (\sigma + \sigma_1)\frac{\alpha}{L\|c\|}$ provided (3.38) holds.*

Proof We consider the evolution of plant error during the time interval $[t_i + \delta_i, t_{i+1} + \delta_{i+1})$ and $i \in \mathbb{Z}_{\geq 0}$. The proof follows here with finding the maximum evolution of $\|e_p\|$ during $[t_i + \delta_i, t_{i+1})$ and $[t_{i+1}, t_{i+1} + \delta_{i+1})$ separately. Note that during $[t_i + \delta_i, t_{i+1})$, the control law (3.34) is applied to the system (3.1)–(3.2). So, from the Theorem 3.1, it follows immediately that the maximum value of $\|e_p\|$ during this time period is $\frac{\sigma\alpha}{L\|c\|}$.

In the next, we obtain the maximum increase in $\|e_p\|$ during the interval $[t_{i+1}, t_{i+1} + \delta_{i+1})$. It may be noted that from Theorem 3.4, if $\delta_i \leq \varepsilon_i$ the error at event-triggering block is bounded by $\|e(t)\| \leq \frac{\sigma_1 \alpha}{L\|c\|}$ for all $t \in [t_{i+1}, t_{i+1} + \delta_{i+1})$ and $i \in \mathbb{Z}_{\geq 0}$. The error in the plant would also increase to this value if it starts from the zero initial condition. But, this error grows from the value $\frac{\sigma\alpha}{L\|c\|}$ during the interval $[t_{i+1}, t_{i+1} + \delta_{i+1})$. So, solving the differential inequality (3.25) for plant error $\|e_p(t)\|$ on replacing $\|e(t)\|$ by $\|e_p(t)\|$ and using initial condition $\frac{\sigma\alpha}{L\|c\|}$ during the time interval $[t_{i+1}, t_{i+1} + \delta_{i+1})$, we obtain

$$
\begin{aligned}
\|e_p(t)\| &\leq \frac{\rho_N(\|x(t_i)\|) + \beta_N}{L}\left(e^{L(t-t_{i+1})} - 1\right) + \frac{\sigma\alpha}{L\|c\|}e^{L(t-t_{i+1})} \\
&< \frac{\rho_N(\|x(t_i)\|) + \beta_N}{L}\left(e^{L\delta_{i+1}} - 1\right) + \frac{\sigma\alpha}{L\|c\|}e^{L\delta_{i+1}} \\
&\leq \frac{\rho_N(\|x(t_i)\|) + \beta_N}{L}\left(e^{L\varepsilon_{i+1}} - 1\right) + \frac{\sigma\alpha}{L\|c\|}e^{L\varepsilon_{i+1}} \\
&\leq \frac{\sigma_1 \alpha}{L\|c\|} + \frac{\sigma\alpha}{L\|c\|}e^{L\varepsilon_{i+1}}.
\end{aligned}
$$

It may be noted that if the delay is assumed to be zero, then $\sigma_1 = 0$ and the error grows to only $\frac{\sigma\alpha}{L\|c\|}$. But, in the case of delay the error in the plant increases to the value given in the above expression. This can be simplified further as follows. Denote the maximum value of $e^{L\varepsilon_{i+1}}$ by m. Then, m becomes maximum when ε_{i+1} does. Since the closed loop is stable, the switching gain satisfies the relation $K > |c^T B|d_0 + \sigma_1 \alpha + m\sigma\alpha$. Now using this in the relation (3.35), the maximum value of ε_{i+1} is obtained when $m = 1$. So, the plant error during the time interval $[t_{i+1}, t_{i+1} + \delta_{i+1})$ remains bounded by $\frac{(\sigma + \sigma_1)\alpha}{L\|c\|}$. Thus, the proof is completed. $\qquad\square$

One important consequence to the above result is given in the following Theorem. This states the stability of the closed loop system with the delayed control action. It is seen that due to delay the plant error increases beyond the threshold value of event-triggering block. So, the event-triggered SMC must be designed such that the practical sliding mode is ensured in the system.

Theorem 3.5 *Consider the system* (3.1)–(3.2). *Let* $\sigma_1 \in (0, 1)$ *and* $\sigma \in (\sigma_1, 1)$ *be given. Also, let* $\alpha > 0$ *be given such that* (3.7) *is satisfied for all time. Then, the practical sliding mode is enforced in the system in finite-time with the control* (3.6) *if*

$$K > \sup_{t \geq 0} \left| c^\top B d(t) \right| + (\sigma + \sigma_1)\alpha. \tag{3.43}$$

Proof The proof follows similar lines of the proof of Theorem 3.1. Consider the Lyapunov function $V = \frac{1}{2}s^2$. Now, differentiating V with respect to time $t \in [t_i + \delta_i, t_{i+1} + \delta_{i+1})$ along the system trajectories of (3.1)–(3.2) and using (3.6), we obtain

$$
\begin{aligned}
\dot{V}(s(t)) &= s(t)\dot{s}(t) \\
&= s(t)\left(c^\top f(x(t)) + B_2 u(t_i) + B_2 d(t)\right) \\
&= s(t)\left(c^\top f(x(t)) - c^\top f(x(t_i)) - K\,\mathrm{sign}s(t_i) + B_2 d(t)\right) \\
&\leq -s(t)K\,\mathrm{sign}s(t_i) + |s(t)|\,\left\|c^\top f(x(t)) - c^\top f(x(t_i))\right\| + |s(t)|\,|B_2|\,d_0 \\
&\leq -s(t)K\,\mathrm{sign}s(t_i) + |s(t)|\,\|c\|\,\|f(x(t)) - f(x(t_i))\| + |s(t)|\,|B_2|\,d_0 \\
&\leq -s(t)K\,\mathrm{sign}s(t_i) + |s(t)|L\,\|c\|\,\|x(t) - x(t_i)\| + |s(t)|\,|B_2|\,d_0 \\
&= -s(t)K\,\mathrm{sign}s(t_i) + |s(t)|L\,\|c\|\,\left\|e_p(t)\right\| + |s(t)|\,|B_2|\,d_0 \tag{3.44}
\end{aligned}
$$

We use here again the same condition, $\mathrm{sign}s(t_i) = \mathrm{sign}s(t)$, before the trajectories reach the sliding manifold. Also, note that $\|e_p(t)\| < (\sigma + \sigma_1)\frac{\alpha}{L}$. Using this and (3.43) in (3.44), we obtain

$$
\begin{aligned}
\dot{V}(s(t)) &\leq -|s(t)|K + |s(t)|(\sigma + \sigma_1)\alpha + |s(t)|\,|B_2|\,d_0 \\
&= -|s(t)|\left(K - (\sigma + \sigma_1)\alpha - |B_2|\,d_0\right) \\
&< -\eta|s(t)| \tag{3.45}
\end{aligned}
$$

for some $\eta > 0$. This shows that the trajectory moves towards the sliding manifold. This process continues for the subsequent triggering intervals as long as $\mathrm{sign}s(t_i) = \mathrm{sign}s(t)$. Eventually, the trajectory hits the manifold in finite-time due to (3.45). However, it is not guaranteed that the trajectory remains on this manifold as the control signal is not applied continuously. So, the trajectory crosses the manifold after hitting it. But, it does not go unbounded due to Proposition 3.2 which is shown below.

We obtain the maximum deviation of sliding trajectory in any time interval $[t_i + \delta_i, t_{i+1} + \delta_{i+1})$. This can be written using (3.7) as

$$
\begin{aligned}
|s(t_i) - s(t)| &= \left| c^\top x(t_i) - c^\top x(t) \right| \\
&\leq \|c\| \, \|e_p(t)\| \\
&< (\sigma + \sigma_1) \frac{\alpha}{L}.
\end{aligned}
$$

The maximum value of practical sliding mode band can be obtained if the triggering takes place when the trajectory just reaches $\mathscr{S}$, i.e. $s(t_i) = 0$, and thus, the band is given as

$$
\Omega_d = \left\{ x \in \mathscr{D} : |s| = \left| c^\top x \right| < (\sigma + \sigma_1) \frac{\alpha}{L} \right\}. \tag{3.46}
$$

This shows that the system trajectory is ultimately bounded in the region Ω_d. Thus, the proof is completed. $\qquad\square$

3.5 Simulation Results

In this section, we present the simulation results pertaining to the above analysis. Consider a second-order nonlinear system

$$
\begin{aligned}
\dot{x}_1 &= x_2 \\
\dot{x}_2 &= x_1 + x_2^2 + u + d.
\end{aligned} \tag{3.47}
$$

The domain of interest for the system is chosen as $\mathscr{D} = \{x \in \mathbb{R}^2 : \|x\|^2 = 9\}$. The Lipschitz constant for the system in this domain $\mathscr{D}$ is selected as $L = 10$. The perturbed disturbance is assumed to be bounded and is taken here as $d = 0.5 \cos t$ and this gives $d_0 = 0.5$. The sliding surface is designed such that reduced order system (upper part of (3.47)) is stable. We select $c^\top = [\,0.5\ 1\,]$, so $s = c^\top x$ denotes sliding variable in continuous-time setting. The SMC for (3.47) can be given as

$$
u = -x_1 - 0.5x_2 - x_2^2 - K \operatorname{sign} s \tag{3.48}
$$

with $K > 0.5$. When the control (3.48) is implemented with event-triggered strategy, it is held constant in the interval $[t_i, t_{i+1})$, i.e. $u(t) = u(t_i)$ for all time $t \in [t_i, t_{i+1})$ and $i \in \mathbb{Z}_{\geq 0}$. The other parameters of interest are chosen as $T_o^\star = 0.0001$, $\sigma = 0.8$. The simulation is run with initial condition $\begin{bmatrix} -1 & 2 \end{bmatrix}^\top$. Here, we study both delay and delay-free case separately.

Fig. 3.2 Performance of event-triggered SMC without delay

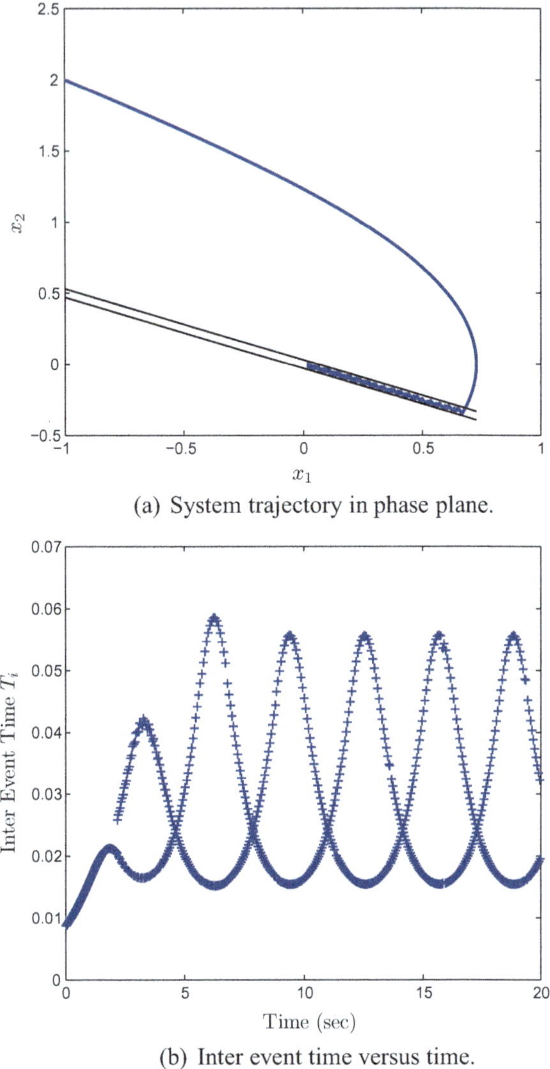

(a) System trajectory in phase plane.

(b) Inter event time versus time.

3.5.1   *Without Delay*

The initial value of $\alpha_{\min}$ selected to be 0.3. Based on Algorithm 1, we arrive at $K = 0.88$. Figures 3.2 and 3.3 show the simulation results of event-triggered SMC for the nonlinear system (3.47). In Fig. 3.2a, it is shown that the practical sliding mode occurs in the system in finite-time. For the selected values of α, the value

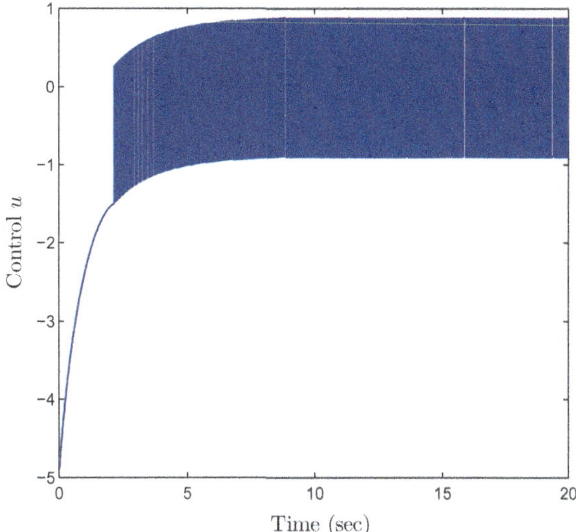

Fig. 3.3 Event-triggered SMC signal without delay

of the sliding mode band is given as 0.03. The sliding trajectories enter the band in finite-time and remain there for all time. This is also true for different values of disturbance bounds. The variation of inter-execution time is shown in Fig. 3.2b. Once the sliding mode is enforced, the inter-execution time is increased as high as 0.06. Due to this, the processor can perform many tasks simultaneously and eventually increases system performance. This has direct effect on the control updates which is shown in Fig. 3.3.

3.5.2 With Delay

Here, we choose $\sigma_1 = 0.2$ and $\sigma = 0.8$. The initial value of $\alpha_{\min}$ selected to be 0.3 and $K^\star$ is chosen as 0.88 according to (3.43). Based on Algorithm 1, we arrive at $K = 0.88$. Clearly $\sigma > \sigma_1$. Then, it ensures that event condition (3.20) is always satisfied due to Theorem 3.4. The performance of system (3.47) with event-triggered SMC is shown in Figs. 3.4 and 3.5. In steady state, the trajectory remains bounded with bound given as $(\sigma + \sigma_1)\frac{\alpha}{L} = 0.03$ which is shown in Fig. 3.4a. We observe that the effects of delay on the inter-execution time which is shown in Fig. 3.4b. However, unlike delay-free case, the inter-execution time intervals are increased due to inherent delay in control updates. The corresponding plot of control signal is shown in Fig. 3.5.

Fig. 3.4 Performance of event-triggered SMC with delay

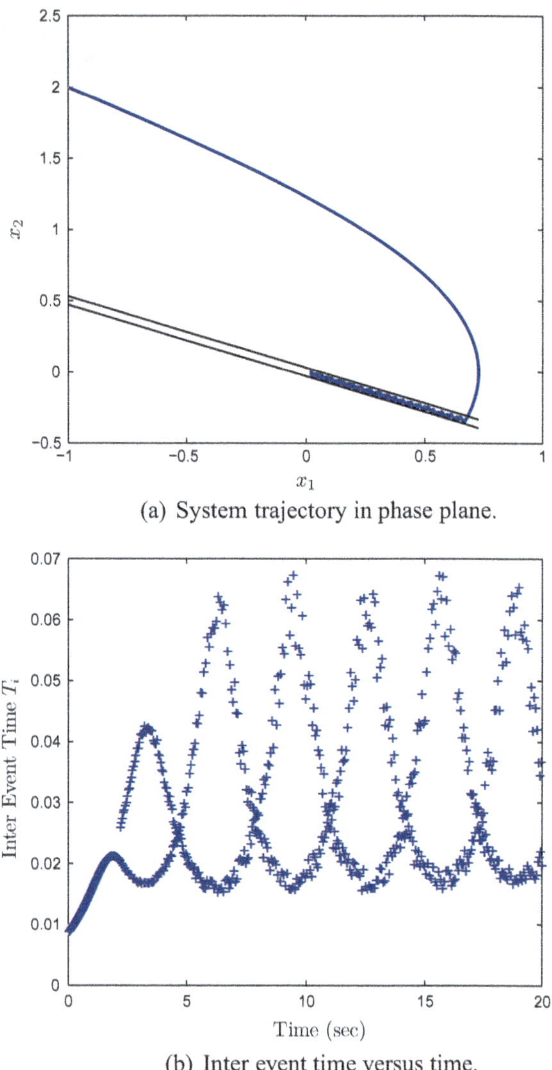

(a) System trajectory in phase plane.

(b) Inter event time versus time.

3.6 Summary

This chapter presents the design of event-triggered SMC for a class of nonlinear systems. First, the design of SMC is introduced in the system in continuous-time domain. Then, a sufficient condition for the existence of practical sliding mode is given in detail for the event-triggered implementation of SMC. The closed loop

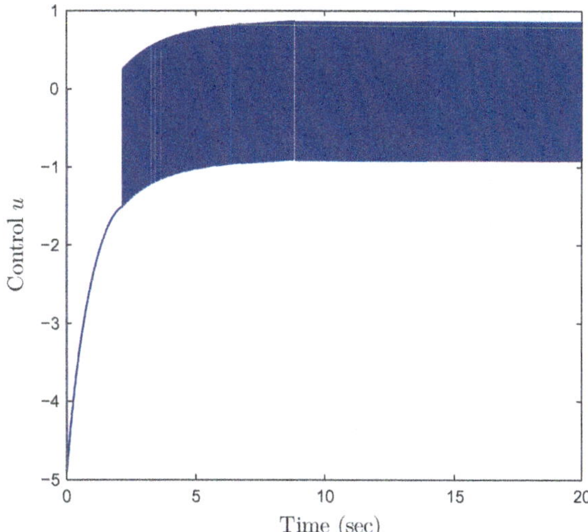

Fig. 3.5 Event-triggered SMC signal with delay

system stability is shown in detail using the proposed triggering mechanism. Moreover, the analysis is extended to delayed control execution at the plant. In this case, the upper bound of the delay is given such that the closed loop system is stable. Finally, simulation results are given to demonstrate the performance of event-triggered SMC.

3.7 Notes and References

The preliminaries on the design of SMC for nonlinear systems can be referred in [1, 3–8]. The event-triggering-based design of SMC for LTI systems can be found in [9, 10]. The more recent developments for event-triggered SMC can be found in [11–13], and for nonlinear systems see [12]. Our approach is based on the preliminary results on event-triggered control presented in [14–20].

References

1. A.F. Filippov, *Differential Equations With Discontinuous Right-Hand Sides* (Kluwer Academic Publishers, Dordrecht, 1988)
2. H. Khalil, *Nonlinear Systems*, 3rd edn. (Prentice-Hall, Upper Saddle River, NJ, 2002)
3. V.I. Utkin, *Sliding Modes and Their Applications in Variable Structure Systems*, Translated from the Russian by A. Parnakh (MIR Publishers, Moscow, 1978)
4. V.I. Utkin, J. Gulnder, J. Shi, *Sliding Mode Control in Electromechanical Systems* (CRC Press, Taylor and Francis Group, 1999)
5. C. Edwards, S.K. Spurgeon, *Sliding Mode Control: Theory and Applications* (CRC Press, Taylor and Francis Group, 1998)

6. H.S. Ramirez, Differential geometric methods in variable-structure control. Int. J. Control **48**(4), 1359–1390 (1988)
7. J.-J. Slotine, S.S. Sastry, Tracking control of nonlinear systems using sliding surfaces with application to robot manipulator. Int. J. Control **38**(2), 465–492 (1983)
8. J.-J. Slotine, Sliding controller design for non-linear systems. Int. J. Control **40**(2), 421–434 (1984)
9. A.K. Behera, B. Bandyopadhyay, Event based robust stabilization of linear systems, in *Proceedings of 40th Annual Conference of the IEEE Industrial Electronics Society, Dallas, USA* (2014), pp. 133–138
10. A. Ferrara, G.P. Incremona, L. Magini, Model-Based Event-Triggered Robust MPC/ISM, in *Proceedings of the 13th European Control Conference, Strausbourg, France* (2014), pp. 2931–2936
11. A.K. Behera, B. Bandyopadhyay, Event based sliding mode control with quantized measurement, in *Proceedings of International Workshop on Recent Advances Sliding Modes, Istanbul, Turkey* (2015), pp. 1–6
12. A.K. Behera, B. Bandyopadhyay, Event-triggered sliding mode control for a class of nonlinear systems. Int. J. Control **89**(9), 1916–1931 (2016)
13. A.K. Behera, B. Bandyopadhyay, Robust sliding mode control: an event-triggering approach. IEEE Trans. Circuits Syst. II Express Briefs **64**(2), 146–150 (2017)
14. K.-E. Årźen, A simple event-based PID controller, in *Proceedings of the 14th IFAC World Congrress*, Beijing, China (1999), pp. 423–428
15. K.J. Åström, B. Bernhardsson, Comparison of Riemann and Lebesgue sampling for first order stochastic systems, in *Proceedings of 41st IEEE Conference on Decision and Control, Las Vegas, USA* (2002), pp. 2011–2016
16. P. Tabuada, Event-triggered real-time scheduling of stabilizing control tasks. IEEE Trans. Autom. Control **52**(9), 1680–1685 (2007)
17. Y.-K. Xu, X.-R. Cao, Lebesgue-sampling-based optimal control problems with time aggregation. IEEE Trans. Autom. Control **56**(5), 1097–1109 (2011)
18. W.P.M.H. Heemels, K.H. Johansson, P. Tabuada, An introduction to event-triggered and self-triggered control, in *Proceedings of 51st IEEE Conference of Decision and Control, Hawai, USA* (2010), pp. 3270–3285
19. P. Tabuada, X. Wang, Preliminary results on state-triggered stabilizing control tasks, in *Proceedings of 45th IEEE Conference on Decision and Control, San Deigo, USA* (2006), pp. 282–287
20. J. Lunze, D. Lehmann, A state-feedback approach to event-based control. Automatica **46**(1), 211–215 (2010)

Chapter 4
Self-Triggered Sliding Mode Control for Linear Systems

In the previous chapters, the event-triggered SMC strategy is discussed for linear and nonlinear systems. The event-triggering strategy is a novel control implementation technique where the control signal is updated to the plant at some aperiodic time instants generated by executing a stabilizing triggering rule. The triggering scheme continuously monitors the state evolution of the system and generates possible triggering instants. In order that sophisticated sensors are deployed to obtain almost continuous measurements which is then used by the event-triggering scheme. For implementing such strategy, a complete hardware is needed. This may not be always desirable in some practical applications.

This chapter, on the other hand, provides an alternative to realize the event-triggering strategy, known as self-triggering. In this strategy, the continuous monitoring of state trajectory is not required to determine the triggering instant. The sampled value of the state at the triggering instant is only required to generate the triggering instant for updating the control signal. So, no dedicated circuit is required in this strategy, and thus the cost of control implementation is reduced compared to event-triggering technique. However, the inter event time in this strategy is decreased due to relaxation of continuous state measurements.

4.1 System Description

Consider a SISO LTI system given by (1.11) and is rewritten as

$$\dot{x} = Ax + B(u + d).$$

© Springer International Publishing AG, part of Springer Nature 2018

B. Bandyopadhyay and A. K. Behera, *Event-Triggered Sliding Mode Control*, Studies in Systems, Decision and Control 139, https://doi.org/10.1007/978-3-319-74219-9_4

We design SMC for the above system and discuss the stability of SMC with event-triggered implementation. Consider the sliding manifold given by (1.14) for the sliding variable $s = c^\top x$. The surface parameter c is designed to ensure stability of sliding motion dynamics. The SMC law that brings the system trajectory to the sliding manifold $\mathcal{S}$ is

$$u = -(c^\top B)^{-1} \left(c^\top A x + K \operatorname{sign} s \right) \tag{4.1}$$

where $K > \sup_{t \geq 0} |c^\top B d(t)|$. When the trajectory reaches the manifold, it moves towards the origin without leaving the manifold. This is called as a sliding mode.

In this book, the discrete implementation of the control law (4.1) is discussed. Mostly, we focus on event-triggered implementation of SMC and is given below as

$$u(t) = -(c^\top B)^{-1} \left(c^\top A x(t_i) + K \operatorname{sign} s(t_i) \right) \tag{4.2}$$

for all $t \in [t_i, t_{i+1})$. In this chapter, we provide a triggering mechanism which generates the triggering instants based on the past sampled information only. Also, we develop the triggering condition such that sliding trajectory remains bounded in the vicinity of the sliding manifold. Here, the proposed self-triggering mechanism is motivated by the event-triggering strategy. That means, the triggering instants in self-triggering mechanism are generated such that the stability condition for the system obtained in the event-triggering strategy is always respected. As a result, the stability of the closed loop system is guaranteed.

Let $\{t_i\}_{i=0}^{\infty}$ be the sequences of triggering instants. In Chap. 2, it is assumed that the control signal is updated at these triggering instants. But, this may not be true in many practical applications. Indeed, there is always some delay associated with the control execution, and sometimes it is not possible to ignore the delay in the analysis if it is large enough. For sufficiently small delays, the analysis of without delay case may be carried out without deteriorating the system performance. Let δ_i be the delay associated with the control execution at the time instant t_i. Then, $\{t_i + \delta_i\}_{i=0}^{\infty}$ represents the sequence of time instants for updating control signals. We define $e(t) = x(t_i) - x(t)$ as the error due to the discrete implementation of control law.

In the following, at first the self-triggering scheme is presented without considering any delay in control execution. Then, the result is extended to the delay case. The detailed design steps are given to guarantee the stability of the self-triggered control system with delayed control task.

4.2 Self-Triggering Scheme Without Delay

It is seen that the event-triggered SMC results practical sliding mode in the system in finite-time. The essential features in event-triggered SMC are reported in Theorem (2.1) (see the relation (2.2)). The switching gain is designed such that it dominates the growth of error, $e(t)$, maintained by the event-triggering scheme. So, as long as the relation

$$\|c\|\|A\|\|e(t)\| < \sigma\alpha, \qquad \sigma \in (0, 1) \tag{4.3}$$

holds due to triggering scheme (2.12), the practical sliding mode is enforced in the system. Then, further it is established that there always exists a minimum positive lower bound for the triggering scheme such that (4.3) holds for all time $t \geq 0$. This is established in Theorem 2.2, and we recall the lower bound for inter-event time as

$$T_i \geq \frac{1}{\|A\|} \ln\left(1 + \sigma \frac{\alpha}{\|c\|(\rho(x(t_i)) + \beta)}\right) =: \Xi(x(t_i), \sigma)$$

where $\rho(x(t_i))$ and β are defined by (2.14) and (2.15), respectively. That means, in the time duration $\Xi(x(t_i), \sigma)$ there is no triggering instants generated, and hence monitoring of event in the time interval $[t_i, t_i + \Xi(x(t_i), \sigma))$ may be avoided. Thus, in delay-free case, the self-triggering scheme is given as

$$t_{i+1} = t_i + \Xi(x(t_i), \sigma). \tag{4.4}$$

From Eq. (2.13), $\Xi(x(t_i), \sigma)$ is always strictly greater than zero. So, positive lower bound for inter-event time is guaranteed. This self-triggering scheme uses sampled state at the triggering instant $x(t_i)$ and estimates the next triggering instant from the relation (4.4). The performance of this triggering scheme is illustrated through a numerical simulation.

Proposition 4.1 *Consider the system (1.11) and the control law (4.1). The closed loop system is stable if the triggering instants are generated by (4.4).*

Proof Follows from above.   □

4.2.1 Simulation Results

Consider a numerical example

$$\dot{x} = \begin{bmatrix} 0 & 1 \\ 4 & 5 \end{bmatrix} x + \begin{bmatrix} 0 \\ 1 \end{bmatrix} (u + 0.5 \sin t).$$

First we will design a SMC for the system. The sliding variable parameter is chosen as $c = [0.5 \; 1]^\top$. It is easy to check that stable sliding motion results from this choice of parameter and also $c^\top B$ is nonsingular. The event parameters α and σ are selected as 0.3 and 0.85, respectively. From the relation (2.3), we see that

Fig. 4.1 Performance of self-triggered SMC for LTI systems

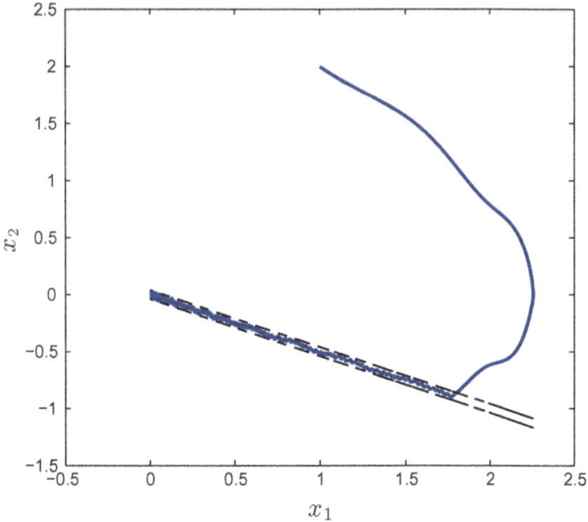

(a) System trajectory in phase plane.

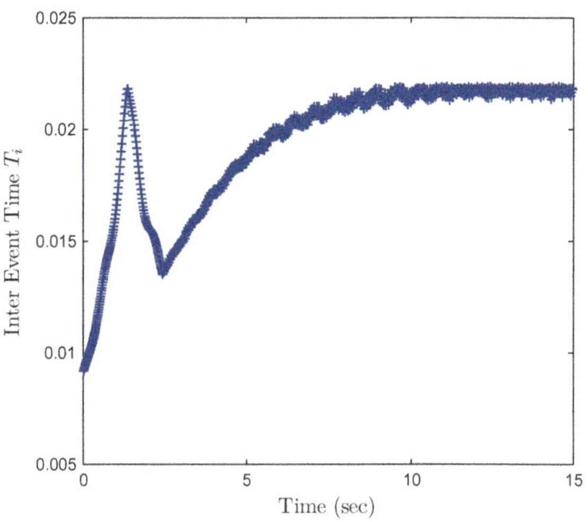

(b) Inter event time versus time.

the switching gain, K, must satisfy $K > 0.8$, and here it is designed as $K = 1$. The event-triggered SMC is thus given as

$$u = -\left([4\ 5.5] \begin{bmatrix} x_1 \\ x_2 \end{bmatrix} + \mathrm{sign}s \right).$$

This control is implemented using self-triggering strategy. The simulation is run in MATLAB with $x(0) = [1\ 2]^\top$. The performance of self-triggered SMC without considering delay is shown in Figs. 4.1 and 4.2. In Fig. 4.1a, the state trajectory in

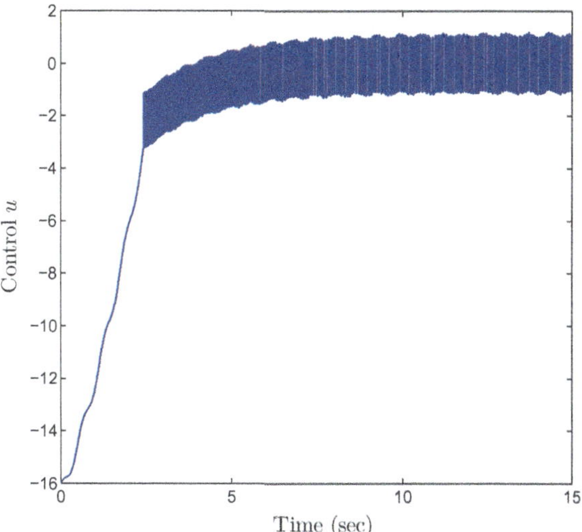

Fig. 4.2 Self-triggered SMC signal for LTI systems

phase plane remains bounded within the practical sliding mode band while the system is in practical sliding mode. The triggering instant is generated by self-triggering rule (4.4). The variation of inter-event time is shown in Fig. 4.1b. The inter-event time is always lower bounded from zero, and thus the stability is assured as claimed in the Proposition 4.1. The control signal is also plotted in Fig. 4.2.

4.3 Self-Triggering Scheme with Delay

In many practical applications, the delays associated with the control execution are not negligible, and so it must be accounted in the analysis. These delays may arise due to control computation, delay due to actuation, etc. The control signal is updated only after the delay is elapsed. If $\{t_i\}_{i=0}^{\infty}$ is the sequence of triggering instants, then $\{t_i + \delta_i\}_{i=0}^{\infty}$ denotes the sequence of control updating instants.

The system dynamics with delay δ_i associated with sampling/triggering instant t_i, can be written as

$$\dot{x}(t) = Ax(t) + Bu(t_{i-1}) + Bd(t) \tag{4.5}$$

$$\dot{x}(t) = Ax(t) + Bu(t_i) + Bd(t) \tag{4.6}$$

for $t \in [t_i, t_i + \delta_i)$ and $t \in [t_i + \delta_i, t_{i+1} + \delta_{i+1})$, respectively. Here, the immediate past control is acting in the system due to the delay, and hence the stability of the system is to be analysed by considering the effects of the delays. In other words, the self-triggering strategy needs to be designed to ensure the stability of ETCS. A detailed analysis on the stability of the system with delay is discussed below.

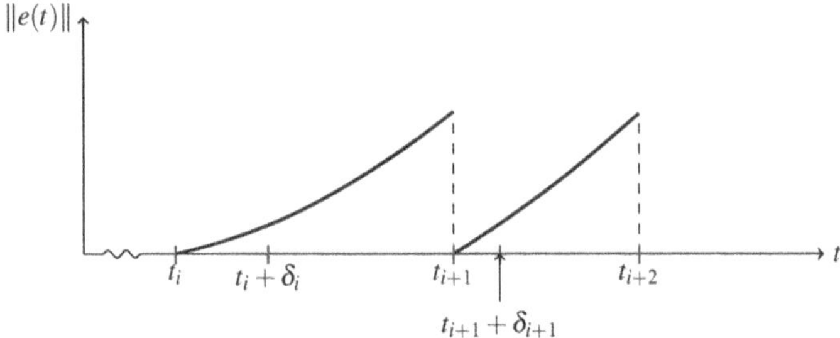

Fig. 4.3 Evolution of $\|e(t)\|$ with time in the self-triggering strategy

Let t_i be the triggering instant generated by the self-triggering strategy. Then, the control is updated or applied to plant at the time instant $t_i + \delta_i$ due to delayed control task as shown in Fig. 4.3. Since the self-triggering strategy is motivated from the event-triggered strategy, the error must be bounded by $\frac{\alpha}{\|c\|\|A\|}$ for all time $t \in [t_i, t_{i+1})$. However, a sufficiently large delay δ_i may lead to the case where triggering instant, t_{i+1} is generated before. This shows that the triggering instant is generated before previous control signal is updated. As a result of this, the control signal gets queued up over a time interval, and eventually the system may become unstable due to not updating the control signal. Summarizing, the error between any consecutive triggering instants must be smaller than $\frac{\alpha}{\|c\|\|A\|}$ for all time to ensure the stability of the system. Following strategy is proposed to guarantee this holds.

This situation can be avoided if $t_i + \delta_i < t_{i+1}$ for all $i \in \mathbb{Z}_{\geq 0}$. We call such triggering sequence as *admissible*. A self-triggering strategy is developed that ensures control signal is always updated before next triggering instant occurs. This needs a careful redesigning of the self-triggering strategy considering the delay into account. Here, we provide a sufficient condition on the delay in terms of its upper bound to design the self-triggering strategy.

The following Lemma gives an upper bound on delay to ensure triggering condition is respected for all time. The idea of the proof is to find the error bound in time interval $[t_i, t_{i+1})$ such that it remains within a pre-specified value.

Lemma 4.1 *Consider the system* (4.5). *Let* $\sigma_1 \in (0, 1)$ *be given any constant. Define the followings as*

$$\Xi_1(x(t_i), x(t_{i-1}), \sigma_1) = \frac{1}{\|A\|} \ln\left(1 + \sigma_1 \frac{\alpha}{\|c\|(\rho_1(x(t_i), x(t_{i-1})) + \beta)}\right), \quad (4.7)$$

and

$$\phi(x(t_i), x(t_{i-1}), t - t_i) = \frac{\rho_1(x(t_i), x(t_{i-1})) + \beta}{\|A\|}\left(e^{\|A\|(t - t_i)} - 1\right) \quad (4.8)$$

where

$$\rho_1(x(t_i), x(t_{i-1})) = \left\| Ax(t_i) - B(c^\top B)^{-1}c^\top Ax(t_{i-1}) \right\|. \tag{4.9}$$

If $\delta_i \leq \Xi_1(x(t_i), x(t_{i-1}), \sigma_1)$ for all $t \in [t_i, t_i + \delta_i)$ and $i \in \mathbb{Z}_{\geq 0}$, then the time evolution of $\|e(t)\|$ satisfies

$$\|e(t)\| < \frac{\sigma_1\alpha}{\|c\|\|A\|} \tag{4.10}$$

for all $t \in [t_i, t_i + \delta_i)$ and $i \in \mathbb{Z}_{\geq 0}$.

Proof Consider $\Gamma_1 = \{t \in [t_i, t_i + \delta_i) : \|e(t)\| = 0\}$. Then, for $t \in [t_i, t_i + \delta_i)\backslash\Gamma_1$, we write

$$
\begin{aligned}
\frac{\mathrm{d}}{\mathrm{d}t}\|e(t)\| &\leq \left\| \frac{\mathrm{d}}{\mathrm{d}t}e(t) \right\| = \left\| \frac{\mathrm{d}}{\mathrm{d}t}x(t) \right\| \\
&= \left\| Ax(t) - B(c^\top B)^{-1}c^\top Ax(t_{i-1}) - B(c^\top B)^{-1}K\,\mathrm{sign}s(t_{i-1}) + Bd(t) \right\| \\
&= \left\| Ax(t_i) - Ae(t) - B(c^\top B)^{-1}c^\top Ax(t_{i-1}) - B(c^\top B)^{-1}K\,\mathrm{sign}s(t_{i-1}) \right. \\
&\qquad \left. + Bd(t) \right\| \\
&\leq \|A\|\|e(t)\| + \|B(c^\top B)^{-1}K\| + \|B\|d_0 + \left\| Ax(t_i) - B(c^\top B)^{-1}c^\top Ax(t_{i-1}) \right\| \\
&= \|A\|\|e(t)\| + \rho_1(x(t_i), x(t_{i-1})) + \beta \tag{4.11}
\end{aligned}
$$

where $\rho_1(x(t_i), x(t_{i-1}))$ and β are defined by (4.9) and (2.15), respectively. Since $\|e(t_i)\| = 0$ at the event-triggering block, the solution to the differential inequality (4.11) can be given as

$$\|e(t)\| \leq \phi(x(t_i), x(t_{i-1}), t - t_i) \tag{4.12}$$

for $t \in [t_i, t_i + \delta_i)$. It can be seen that $\phi(x(t_i), x(t_{i-1}), t - t_i)$ is a monotonic function with $t - t_i$. So, we can write immediately

$$
\begin{aligned}
\|e(t)\| &\leq \phi(x(t_i), x(t_{i-1}), t - t_i) \\
&< \phi(x(t_i), x(t_{i-1}), \delta_i). \tag{4.13}
\end{aligned}
$$

Since $\delta_i \leq \Xi_1(x(t_i), x(t_{i-1}), \sigma_1)$ holds, the above relation (4.13) can be simplified using (4.7) to

$$\|e(t)\| < \frac{\sigma_1\alpha}{\|c\|\|A\|} \tag{4.14}$$

for all time $t \in [t_i, t_i + \delta_i)$. This completes the proof. $\square$

It is seen that in the time interval $[t_i, t_i + \delta_i)$ the error $\|e(t)\|$ grows at most to $\frac{\sigma_1\alpha}{\|c\|\|A\|}$. In the next Lemma, we find the evolution of error in the time interval $t \in [t_i + \delta_i, t_{i+1})$. The main purpose of the following Lemma is to show the error always remain bounded during the interval $[t_i + \delta_i, t_{i+1})$ under the assumption of delay in Lemma 4.1.

Lemma 4.2 *Consider the system (4.6). Let $\sigma_2 \in (\sigma_1, 1)$ be given. Define*

$$\Xi_2(x(t_i), x(t_{i-1}), \sigma_2) = \frac{1}{\|A\|} \ln\left(1 + \frac{\sigma_2\alpha - \sigma_1\alpha}{\|c\|\left(\rho(\|x(t_i)\|) + \beta\right) + \sigma_1\alpha}\right). \quad (4.15)$$

If $\delta_i \leq \Xi_1(x(t_i), x(t_{i-1}), \sigma_1)$, then $\Xi_2(x(t_i), x(t_{i-1}), \sigma_2) > 0$ and $\|e(t)\| < \frac{\sigma_2\alpha}{\|c\|\|A\|}$ for all $t \in [t_i + \delta_i, t_i + \delta_i + \Xi_2(x(t_i), x(t_{i-1}), \sigma_2))$.

Proof The first part of the proof that $\Xi_2(x(t_i), x(t_{i-1}), \sigma_2) > 0$ is immediate as $\sigma_2 > \sigma_1$. To prove the second part, we obtain time derivative of $\|e(t)\|$ for all time in the interval $[t_i + \delta_i, t_i + \delta_i + \Xi_2(x(t_i), x(t_{i-1}), \sigma_2))$. Let $\Gamma_2 = \{t \in [t_i + \delta_i, t_i + \delta_i + \Xi_2(x(t_i), x(t_{i-1}), \sigma_2)) : \|e(t)\| = 0\}$. Then, for all $t \in [t_i + \delta_i, t_i + \delta_i + \Xi_2(x(t_i), x(t_{i-1}), \sigma_2))\backslash\Gamma_2$, we have

$$\frac{\mathrm{d}}{\mathrm{d}t}\|e(t)\| \leq \left\|\frac{\mathrm{d}}{\mathrm{d}t}e(t)\right\| = \left\|\frac{\mathrm{d}}{\mathrm{d}t}x(t)\right\|$$

$$= \left\|Ax(t) - B(c^\top B)^{-1}c^\top Ax(t_i) - B(c^\top B)^{-1}K\,\mathrm{signs}(t_i) + Bd(t)\right\|$$

$$= \left\|Ax(t_i) - Ae(t) - B(c^\top B)^{-1}c^\top Ax(t_i) - B(c^\top B)^{-1}K\,\mathrm{signs}(t_i) + Bd(t)\right\|$$

$$\leq \|A\|\|e(t)\| + \left\|A - B(c^\top B)^{-1}c^\top A\right\|\|x(t_i)\| + \|B(c^\top B)^{-1}K\| + \|B\|d_0$$

$$= \|A\|\|e(t)\| + \rho(\|x(t_i)\|) + \beta \qquad (4.16)$$

where $\rho(\|x(t_i)\|)$ and β are defined by (2.14) and (2.15), respectively. Since $\delta_i \leq \Xi_1(x(t_i), x(t_{i-1}), \sigma_1)$, the error $\|e(t)\|$ remains bounded by $\frac{\sigma_1\alpha}{\|c\|\|A\|}$ for all time $t \in [t_i, t_i + \delta_i)$. Hence, the solution to the above differential inequality (4.16) can be obtained with (4.14) as initial condition,

$$\|e(t)\| \leq \frac{\rho(\|x(t_i)\|) + \beta}{\|A\|}\left(e^{\|A\|(t - t_i - \delta_i)} - 1\right) + \frac{\sigma_1\alpha}{\|c\|\|A\|}e^{\|A\|(t - t_i - \delta_i)} \qquad (4.17)$$

for all time $t \in [t_i + \delta_i, t_i + \delta_i + \Xi_2(x(t_i), x(t_{i-1}), \sigma_2))$. Now using (4.15) in the relation (4.17) and then with some simple calculation, one arrives at

$$\|e(t)\| \leq \frac{\|c\|\left(\rho(\|x(t_i)\|) + \beta\right) + \sigma_1\alpha}{\|c\|\|A\|}e^{\|A\|(t - t_i - \delta_i)} - \frac{\rho(\|x(t_i)\|) + \beta}{\|A\|}$$

$$< \frac{\|c\|\left(\rho(\|x(t_i)\|) + \beta\right) + \sigma_1\alpha}{\|c\|\|A\|}e^{\|A\|\Xi_2(x(t_i), x(t_{i-1}), \sigma_2)} - \frac{\rho(\|x(t_i)\|) + \beta}{\|A\|}$$

$$< \frac{\sigma_2\alpha}{\|c\|\|A\|} \qquad (4.18)$$

for all $t \in [t_i + \delta_i, t_i + \delta_i + \Xi_2(x(t_i), x(t_{i-1}), \sigma_2))$. Thus, the proof is completed. $\square$

The preceding Lemmas show that the error always remain bounded if the delay is bounded by some known function. These results will be used to show the stability of the ETCS. A sufficient condition is proposed such that the triggering instants are generated only after the previous control task is executed, i.e. $t_{i+1} > t_i + \delta_i$ for all $i \in \mathbb{Z}_{\geq 0}$. Hence, the stability of the closed loop system is ensured.

Theorem 4.1 *Consider the system* (4.5) *and* (4.6). *Let* $\sigma_1 \in (0, 1)$ *and* $\sigma \in (\sigma_1, 1)$ *be given. Let* δ_i *be the delay associated with the triggering instant* t_i. *Then, the self-triggering strategy given by*

$$t_{i+1} = t_i + \delta_i + \Xi_2(x(t_i), x(t_{i-1}), \sigma) \tag{4.19}$$

generates an admissible triggering sequence $\{t_i\}_{i=0}^{\infty}$ *if* $\delta_i \leq \Xi_1(x(t_i), x(t_{i-1}), \sigma_1)$.

Proof The proof follows from the results of Lemmas 4.1 and 4.2. Note that since $\delta_i \leq \Xi_1(x(t_i), x(t_{i-1}), \sigma_1)$, the error $\|e(t)\|$ grows at most by $\frac{\sigma_1 \alpha}{\|c\|\|A\|}$ due to Lemma 4.1 during the time interval $[t_i, t_i + \delta_i)$. First, observe that $\Xi_2(x(t_i), x(t_{i-1}), \sigma) > 0$ due to $\sigma > \sigma_1$. Indeed, there exists a positive constant Ξ_0 such that $\Xi_2(x(t_i), x(t_{i-1}), \sigma) > \Xi_0$ for all time. This implies that Zeno execution of triggering sequence is avoided, and hence $t_{i+1} > t_i + \delta_i$. Since $\delta_i \leq \Xi_1(x(t_i), x(t_{i-1}), \sigma_1)$, the Lemma 4.2 can be applied. So, it is immediate from Lemma 4.2 that $\|e(t)\| < \frac{\sigma \alpha}{\|c\|\|A\|}$ for all time $t \in [t_i + \delta_i, t_i + \delta_i + \Xi_2(x(t_i), x(t_{i-1}), \sigma))$. Combining all these, we conclude that $\|e(t)\| < \frac{\sigma \alpha}{\|c\|\|A\|}$ for all time $t \in [t_i, t_{i+1})$. Thus, the triggering instant is always generated only after previous control task is executed. This completes the proof. □

Remark 4.1 The self-triggering strategy proposed in (4.19) ensures that the triggering instant is determined from the past sampled values of states and the delay associated with the previous triggering instant. It can be seen that the delay, δ_i, is known once this previous control task is over. So, the self-triggering instant is obtained as soon as delayed control task is finished.

Remark 4.2 At initial time, the self-triggering instant is chosen as $t_0 = 0$. The next self-triggering instant, i.e. t_{i+1} for $i = 0$, is obtained with $x(t_{-1}) = 0$. However, the subsequent self-triggering instants are obtained using the relation (4.19) with all parameters carrying their usual meanings.

4.3.1 Design of Self-Triggered Sliding Mode Control

In the case without delay, the error between two consecutive triggering instants is same as the error in the plant dynamics. So, the error in the plant always remains bounded by $\frac{\sigma \alpha}{\|c\|\|A\|}$ for all time. The self-triggering strategy is developed to ensure the stability. The control law (2.1) with switching gain (2.3) is enough to enforce practical sliding mode in the system. However, it is not necessarily true if there is delay associated with the control task.

In case of delayed control task, the control signal is not applied to the plant at the triggering instant. So, the previous control signal is still acting on the system as given in (4.5) and the plant evolves continuously during this period. As a result, the error in the plant increases until the control signal is applied. In other words, the error in plant does not remain bounded by $\frac{\sigma\alpha}{\|c\|\|A\|}$. Since there is no update in control signal in the interval $[t_i + \delta_i, t_{i+1} + \delta_{i+1})$, error in the plant may increase beyond the bound $\frac{\alpha}{\|c\|\|A\|}$. Denote plant error by $e_p(t) = x(t_i) - x(t)$ for time $t \in [t_i + \delta_i, t_{i+1} + \delta_{i+1})$. For simplifying the analysis, assume that $\delta_0 = 0$, i.e. the control is applied to the plant initially at t_0.

The following Proposition gives the error bound in the plant for all time when the delayed control is applied.

Proposition 4.2 *Consider the system* (4.5) *and* (4.6). *If* $\|e(t)\| < \frac{\sigma\alpha}{\|c\|\|A\|}$ *for all time, then the error in the plant always remain bounded by* $(\sigma + \sigma_1)\frac{\alpha}{\|c\|\|A\|}$, *i.e.* $\|e_p(t)\| < (\sigma + \sigma_1)\frac{\alpha}{\|c\|\|A\|}$.

Proof Note that control $u(t_i)$ is acting on the plant during the time interval $[t_i + \eta_i, t_{i+1} + \eta_{i+1})$. So, the error in the plant grows to a maximum value during this interval only. In the interval $[t_i + \eta_i, t_{i+1})$, error increases to at most $\frac{\sigma\alpha}{\|c\|\|A\|}$. This is a direct consequence of Theorem 4.1. Similarly, during $[t_{i+1}, t_{i+1} + \eta_{i+1})$ it can be shown in the similar manner as in Proposition 3.2 that the error in the plant $\|e_p\|$ grows from $\frac{\sigma\alpha}{\|c\|\|A\|}$ to $(\sigma + \sigma_1)\frac{\alpha}{\|c\|\|A\|}$. This completes the proof. $\qquad\square$

Now, we discuss the stability of the closed loop system with the delayed control execution. The stability of practical sliding mode is shown below by considering the error evolution of plant during the interval $[t_i + \delta_i, t_{i+1} + \delta_{i+1})$. The following Theorem gives a sufficient condition for the existence of practical sliding mode.

Theorem 4.2 *Consider the system* (4.6) *and the control law* (4.2). *Let* $\alpha > 0$ *such that* (2.2) *holds for all time. Then, self-triggered SMC* (4.2) *with self-triggering strategy given by* (4.19) *ensures practical sliding mode in the system if*

$$K > \sup_{t \geq 0} \left|c^\top B d(t)\right| + (\sigma + \sigma_1)\alpha. \tag{4.20}$$

Proof This follows the similar lines of Theorem 2.1. Consider the Lyapunov function $V = \frac{1}{2}s^2$. Then, for all time $t \in [t_i + \delta_i, t_{i+1} + \delta_{i+1})$, the time derivative of V along system trajectory is

$$\begin{aligned}
\dot{V}(s(t)) &= s(t)\dot{s}(t) \\
&= s(t)\left(c^\top A x(t) + c^\top B u(t_i) + c^\top B d(t)\right) \\
&= s(t)\left(-c^\top A e_p(t) - K\,\mathrm{sign}s(t_i) + c^\top B d(t)\right).
\end{aligned} \tag{4.21}$$

Note that unless the sliding manifold $\mathscr{S}$ is reached, $\mathrm{sign}s(t_i) = \mathrm{sign}s(t)$ holds. So, using this and Proposition 4.2, we obtain

$$\dot{V}(s(t)) = -s(t)c^{\mathsf{T}}Ae_p(t) - |s(t)|K + s(t)c^{\mathsf{T}}Bd(t)$$
$$< |s(t)|(\sigma + \sigma_1)\alpha - |s(t)|K + |s(t)|\left|c^{\mathsf{T}}B\right|d_0$$
$$= -|s(t)|\left(K - (\sigma + \sigma_1)\alpha - \left|c^{\mathsf{T}}B\right|d_0\right). \tag{4.22}$$

Now, using (4.20) in (4.22), we have

$$\dot{V}(s) < -\eta|s| \tag{4.23}$$

for some $\eta > 0$. Thus, the system trajectory moves towards the sliding manifold in finite-time. However, it does not slide along $\mathscr{S}$ but remains bounded in the vicinity of $\mathscr{S}$. This bound can be obtained by finding the maximum deviation of sliding trajectory in the interval $[t_i + \delta_i, t_{i+1} + \delta_{i+1})$. Thus,

$$|s(t_i) - s(t)| = \left|c^{\mathsf{T}}x(t_i) - c^{\mathsf{T}}x(t)\right|$$
$$\leq \|c\|\|e_p(t)\|$$
$$< (\sigma + \sigma_1)\frac{\alpha}{\|c\|\|A\|}.$$

However, the maximum bound in the vicinity can be obtained by setting $s(t_i) = 0$. This is given below as

$$\left\{x \in \mathbb{R}^n : |s| = \left|c^{\mathsf{T}}x\right| < (\sigma + \sigma_1)\frac{\alpha}{\|c\|\|A\|}\right\}.$$

This shows that the trajectory is bounded by the event parameter and thus the proof is completed. $\qquad\square$

The self-triggered SMC with delay which is given above differs from that of without delay. In fact, when there is no delay it coincides with self-triggered design of SMC with delay free case. Due to delay, the switching gain is designed by taking into account the effects of the delay. The design steps of self-triggered SMC with delay are given below.

Design steps of self-triggered SMC with delay are encountered in control execution and implementation.

i. Set any desired α. Fix σ_1 for the upper bound on delay and choose any value of σ such that $\sigma \in (\sigma_1, 1)$.
ii. Select the switching gain K using the relation (4.20).
iii. Then, all other parameter can be designed as per standard technique applicable to SMC.

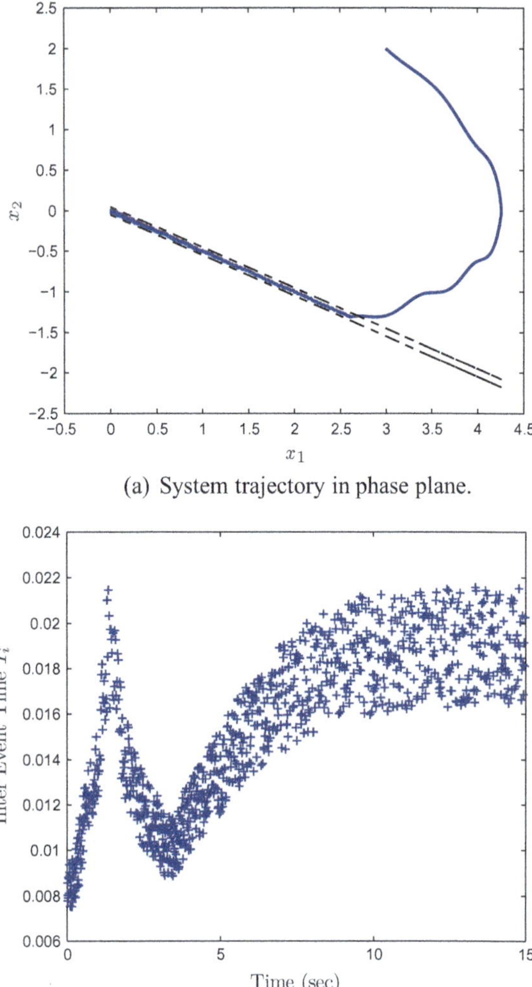

Fig. 4.4 Performance of self-triggered SMC with delay for LTI systems

(a) System trajectory in phase plane.

(b) Inter event time versus time.

4.3.2   Simulation Results

We consider the same example as given in Sect. 4.2.1. So, all the design parameters remain same except those are affected by delay. Here, $\sigma_1 = 0.2$ and $\sigma = 0.85$. The event parameter α is taken as 0.3. The value of K is selected such that it is greater than $(\sigma_1 + \sigma)\alpha + 0.5 = 0.815$, and here in this we choose $K = 1$. With this choice of parameter, the simulation is carried out for self-triggered SMC.

The performance of the self-triggered SMC is shown Figs. 4.4 and 4.5. The system trajectory reaches the sliding manifold in finite-time as shown in Fig. 4.4a and remains

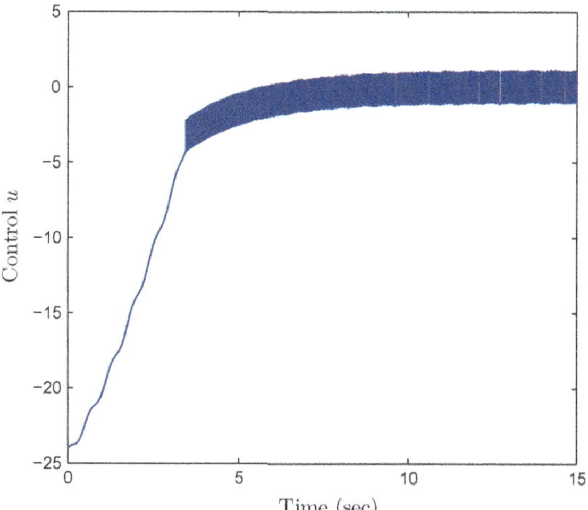

Fig. 4.5 Self-triggered SMC signal with taking delay into account for LTI systems

bounded within the band of size 0.0488. This is in accordance with the result obtained in Theorem 4.2. The variation of inter-event time is shown in Fig. 4.4b. Once the practical sliding mode occurs in the system, the inter-event time is increased to almost double of initial sampling period. It increases to a higher value as large as 0.026. The control signal is also plotted for the system as shown in Fig. 4.5. Overall, it is concluded that similar results are seen with the case of without delay.

4.4 Summary

In the chapter, self-triggered SMC is presented for linear systems. This design is aimed at avoiding the continuous monitoring of the state trajectory for generating possible triggering instant. Rather, the triggering instants are generated from the previous sampled state value at the triggering instants. However, due to this the inter-event time is reduced compared to event-triggered SMC. Nevertheless, the significant benefits are obtained through this self-triggering strategy. Moreover, in this chapter, a more realistic control implementation with delay, due to control computation, is also addressed. A complete analysis of self-triggered SMC is given for the linear systems. Numerical simulation results are given to demonstrate the theoretical developments of this chapter.

4.5 Notes and References

The preliminaries on SMC can be found in [1–5]. The event-triggered control and event-triggered SMC for LTI systems can be found in [6–16]. The deign of self-triggered control strategies are presented in [17–22]. The self-triggered design of SMC can be seen in [14].

References

1. V.I. Utkin, Variable structure systems with sliding modes. IEEE Trans. Autom. Control **22**(2), 212–222 (1977)
2. V.I. Utkin, *Sliding Modes and Their Applications in Variable Structure Systems*, Translated from the Russian by A. Parnakh (Moscow, MIR Publishers, 1978)
3. V.I. Utkin, J. Gulnder, J. Shi, *Sliding Mode Control in Electromechanical Systems* (CRC Press, Taylor and Francis Group, 1999)
4. C. Edwards, S.K. Spurgeon, *Sliding Mode Control: Theory and Applications* (CRC Press, Taylor and Francis Group, 1998)
5. A.F. Filippov, *Differential Equations With Discontinuous Right-Hand Sides* (Kluwer Academic Publishers, Dordrecht, 1988)
6. K.-E. Årźen, A simple event-based PID controller, in *Proceedings of 14th IFAC World Congress, Beijing, China* (1999), pp. 423–428
7. K.J. Åström, B. Bernhardsson, Comparison of Riemann and Lebesgue sampling for first order stochastic systems, in *Proceedings of 41st IEEE Conference on Decision and Control, Las Vegas, USA* (2002), pp. 2011–2016
8. P. Tabuada, Event-triggered real-time scheduling of stabilizing control tasks. IEEE Trans. Autom. Control **52**(9), 1680–1685 (2007)
9. W.P.M.H. Heemels, K.H. Johansson, P. Tabuada, An introduction to event-triggered and self-triggered control, in *Proceedings of 51st IEEE Conference of Decision and Control, Hawai, USA* (2010), pp. 3270–3285
10. A.K. Behera, B. Bandyopadhyay, Event based robust stabilization of linear systems, in *Proceedings of 40th Annual Conference of the IEEE Industrial Electronics Society, Dallas, USA* (2014), pp. 133–138
11. A. Ferrara, G.P. Incremona, L. Magini, Model-Based Event-Triggered Robust MPC/ISM, in *Proceedings of 13th European Control Conference, Strausbourg, France* (2014), pp. 2931–2936
12. A.K. Behera, B. Bandyopadhyay, Event based sliding mode control with quantized measurement, in *Proceedings of International Workshop on Recent Advances Sliding Modes, Istanbul, Turkey* (2015), pp. 1–6
13. A.K. Behera, B. Bandyopadhyay, N. Xavier, S. Kamal, Event-triggered sliding mode control for robust stabilization of linear multivarible systems, in *Recent Advances in Sliding Modes: From Control to Intelligent Mechatronics*, ed. by X. Yu, M. Ö. Efe. Studies in Systems, Decision and Control, vol. 24 (Springer International Publishing, Cham, 2015), pp. 155–175
14. A.K. Behera, B. Bandyopadhyay, Self-triggering-based sliding-mode control for linear systems. IET Control Theory Appl. **9**(17), 2541–2547 (2015)
15. A.K. Behera, B. Bandyopadhyay, Event-triggered sliding mode control for a class of nonlinear systems. Int. J. Control **89**(9), 1916–1931 (2016)
16. A.K. Behera, B. Bandyopadhyay, Robust sliding mode control: an event-triggering approach. IEEE Trans. Circuits Syst-II Express Briefs **64**(2), 146–150 (2017)

17. M. Velasco, J. Fuertes, P. Marti, The self triggered task model for real-time control systems, in *Proceedings of 24th IEEE Real-Time Systems Symposium* (Work Progress Track, Cancun, Mexico, 2003), pp. 67–70
18. X. Wang, M.D. Lemmon, Self-triggered feedback control systems with finite-gain $\mathscr{L}_2$ stability. IEEE Trans. Autom. Control **54**(3), 452–467 (2009)
19. X. Wang, M.D. Lemmon, Self-triggering under state-independent disturbances. IEEE Trans. Autom. Control **55**(6), 1494–1500 (2010)
20. A. Anta, P. Tabuada, To sample or not to sample: self-triggered control for nonlinear systems. IEEE Trans. Autom. Control **55**(9), 2030–2042 (2010)
21. J. Almieda, C. Silvestre, A.M. Pascoal, Self-triggered output feedback controller of linear plants in the presence of unkonwn disturbances. IEEE Trans. Autom. Control **59**(11), 3040–3045 (2014)
22. M. Mazo Jr., A. Anta, P. Tabuada, An ISS self-triggered implementation of linear controllers. Automatica **46**(8), 1310–1314 (2010)

Chapter 5
Discrete Event-Triggered Sliding Mode Control for Linear Systems

The event-triggering strategy has become a popular control implementation technique in these days. To realize this technique, the continuous state measurements are needed to generate the possible triggering instant. This may not be economical in practical applications due to the need of sophisticated sensors and for the computational burden. To avoid such problems, many techniques have been proposed which do not require the continuous measurements such as self-triggering strategy. In a similar spirit of this, a new approach is developed that is based on discrete-time model of the continuous-time system. Since event-triggering strategy is based on the discrete-time model, the continuous evaluation of event-triggering is avoided. Rather, triggering rule is evaluated at periodic intervals and control is updated whenever it is violated only at these periodic instants.

Moreover, in practical systems the full state information may not be available. So, the control may be designed using the output information only. This may be realized using a dynamical observer that estimates the states. However, here, a multirate output feedback-based strategy is used for realizing the output feedback-based discrete event-triggered SMC. One of the popular multirate output feedback techniques, known as fast output sampling (FOS), is used where the outputs are sampled at a faster rate than the control input is applied to the plant. A brief introduction to this design technique is also presented in this chapter. Then, the design of discrete event-triggered SMC is discussed using the multirate output feedback.

5.1 System Description

Consider a SISO dynamical system

$$\dot{x} = Ax + B(u + d) \quad \text{and} \quad y = Cx. \tag{5.1}$$

© Springer International Publishing AG, part of Springer Nature 2018

B. Bandyopadhyay and A. K. Behera, *Event-Triggered Sliding Mode Control*, Studies in Systems, Decision and Control 139, https://doi.org/10.1007/978-3-319-74219-9_5

Note that, this is the same system as given in (1.11). Here, $y \in \mathbb{R}$ is the output of the system (5.1) and C is a constant matrix of appropriate dimension. Since we are interested in design of discrete event-triggering strategy for this system, we obtain the discrete-time model of the system (1.1). Assumption 1.1 on $d(t)$ holds for the system (1.1). The popular ZOH discretization is used to obtain the discrete-time model. The following assumption is made on the disturbance for the discrete-time model.

Assumption 5.1 The disturbance $d(t)$ is slowly varying compared to the sampling rate of the output measurements. So, it is reasonable to assume that the disturbance remains almost constant in between any two consecutive sampling instants. Further, it is also bounded for all time, and hence, it can be written for discrete-time instants as $\sup_{k \in \mathbb{Z}_{\geq 0}} |d(k)| \leq d_0$.

The above assumption is indeed required in the design of discrete-time SMC, otherwise the disturbance becomes mismatched. Using these facts, the ZOH discretization model can be obtained as

$$x(k+1) = \overline{\Phi}_\tau x(k) + \overline{\Gamma}_\tau (u(k) + d(k)) \tag{5.2}$$
$$y(k) = C x(k) \tag{5.3}$$

where Φ_τ and Γ_τ are system and input matrices, respectively, for discrete-time systems that depend on the sampling period. This is given below

$$\overline{\Phi}_\tau = e^{A\tau} \quad \text{and} \quad \overline{\Gamma}_\tau = \int_0^\tau e^{As}\,ds\,B. \tag{5.4}$$

Note that in the discrete representation $x(k)$ actually denotes $x(k\tau)$ and this is true for all other variables also. For the sake of brevity, here τ is omitted if it is clear from the context. However, it is written explicitly whenever it is required. Another important observation is that the discrete-time model is not in regular form even if the continuous-time is. But, this can be transformed into regular form with some suitable transformation matrix. Let T be a nonsingular transformation matrix such that it transforms the state x to a new state space domain given by $z = Tx$. Thus, the transformed system can be obtained from (5.2) as

$$z(k+1) = \Phi_\tau z(k) + \Gamma_\tau (u(k) + d(k)) \tag{5.5}$$
$$y(k) = C_\tau z(k) \tag{5.6}$$

where $\Phi_\tau = T\overline{\Phi}_\tau T^{-1}$, $\Gamma_\tau = T\overline{\Gamma}_\tau$ and $C_\tau = CT^{-1}$. Since the system (5.5) is in regular form, it can be rewritten in the following decomposed form as

$$z_1(k+1) = \Phi_{\tau_{11}} z_1(k) + \Phi_{\tau_{12}} z_2(k) \tag{5.7}$$
$$z_2(k+1) = \Phi_{\tau_{21}} z_1(k) + \Phi_{\tau_{22}} z_2(k) + \Gamma_{\tau_2}(u(k) + d(k)) \tag{5.8}$$

where $z_1 \in \mathbb{R}^{n-1}$ and $z_2 \in \mathbb{R}$. The other matrices are of appropriate compatible dimensions. Now, throughout this chapter we concentrate on the design of event-triggered DTSM using (5.5).

5.2 Discrete-Time Sliding Mode

Consider the system dynamics (5.5). Design sliding variable $s = c^\top z$ where c is a column vector of appropriate dimension. The sliding manifold is defined as

$$\mathscr{S} := \left\{ z \in \mathbb{R}^n : s = c^\top z = 0 \right\}. \tag{5.9}$$

Like continuous-time system, the objective is to design a discrete-time control law that brings the trajectory to the sliding manifold and maintain the trajectory on this manifold. This is known as DTSM. While sliding, the sliding dynamics must ensure the system stability. This can be done by designing $c = [c_1^\top \ 1]^\top$ such that $\Phi_{\tau_{11}} - \Phi_{\tau_{12}} c_1^\top$ is Schur stable, i.e., all the eigenvalues of this matrix are located within the unit disk.

It is very well known that in DTSM, exact sliding mode is not possible due to inherent discrete control signal. As a result, the system trajectory does not slide on the manifold $\mathscr{S}$, but remains bounded in the vicinity of $\mathscr{S}$. This is often known as QSM band in literature. Like continuous-time system, the DTSM control, that yields QSM, may contain switching term. However, it is not necessary for switching to appear in the control law. The control may be designed without any switching term that also ensures quasi-sliding mode. There is a difference in QSM in both types of control law. In the former, the sliding trajectory crosses and recrosses the sliding manifold in every successive sampling instant while the remaining within the QSM band. On the other hand, in switching-free type of QSM the sliding trajectory only remains within the QSM band. Here, a formal definition of QSM is given below that is also appearing in literature.

Definition 5.1 *(Quasi-Sliding Mode)* The system (5.5) is said to be in QSM if for some $\varepsilon > 0$ there exists a $\bar{k} \geq 0$ such that the sliding trajectory $s(k) = c^\top z(k)$ satisfies

$$|s(k)| \leq \varepsilon$$

for all $k \geq \bar{k}$. The constant $\varepsilon (> 0)$ is the size of QSM band.

It is seen from the above notion of QSM that the sliding trajectory does not cross and recross the manifold at every sampling instant. There has been reaching laws proposed in literature in DTSM. One of the popular reaching laws that achieves QSM is $s(k + 1) = 0$ for some $k \geq \bar{k}$. If $\bar{k} = 1$, then the sliding trajectory reaches the sliding manifold in one discrete step. This demands a large control effort to bring the trajectory to the sliding manifold. Indeed, the DTSM control can be designed to bring the trajectory to sliding manifold in some desired finite number of discrete steps. Bartoszewicz proposed a reaching law to achieve this objective. The control law derived using this reaching law ensures finite-time reachability to the sliding manifold.

5.2.1 Bartoszewicz's Reaching Law

In this section, we shall briefly review Bartoszewicz's reaching law that is used for designing DTSM control law. The reaching law proposed by Bartoszewicz [10] is given as

$$s(k+1) = \tilde{d}(k) + s_d(k+1) \tag{5.10}$$

where $\tilde{d}(k)$ is an uncertain variable to be defined later with $\sup_{k \in \mathbb{Z}_{\geq 0}} |\tilde{d}(k)| \leq \tilde{d}_0$ and $s_d(k)$ is *a priori* known function such that followings hold,

- If $|s(0)| > 2\tilde{d}_0$, then

$$s_d(0) = s(0)$$
$$s_d(k) \cdot s(0) \geq 0, \quad \text{for any } k \geq 0$$
$$s_d(k) = 0, \quad \text{for any } k \geq k^\star$$
$$|s_d(k+1)| < |s_d(k)| - 2\tilde{d}_0, \quad \text{for any } k < k^\star.$$

- Otherwise, i.e., if $|s(0)| \leq 2\tilde{d}_0$, then $s_d(k) = 0$ for any $k \geq 0$.

If $|s(0)| > 2\tilde{d}_0$, the sliding variable $s(k)$ decreases in each discrete step by an amount at least equal to or more than $2\tilde{d}_0$. The constant $k^\star$ is a positive integer which is used for defining the rate of convergence of $s(k)$ to QSM band in $k^\star$ number of steps. The value of $k^\star$ can be chosen to meet the design constraint and rate of convergence to the manifold $\mathscr{S}$ using the function $s_d(k)$. One such choice of $s_d(k)$ is given as

$$s_d(k) = \frac{k^\star - k}{k^\star} s(0), \quad \text{with} \quad k^\star < \frac{|s(0)|}{2\tilde{d}_0}.$$

This is essential to satisfy the condition $|s_d(k+1)| < |s_d(k)| - 2\tilde{d}_0$ for all $k \in [0, k^\star)$. This can be seen for all $k \in [0, k^\star)$ as

$$|s_d(k+1)| - |s_d(k)| = \frac{k^\star - k - 1}{k^\star}|s(0)| - \frac{k^\star - k}{k^\star}|s(0)|$$

$$= -\frac{1}{k^\star}|s(0)|$$

$$< -2\tilde{d}_0.$$

5.2.2 Design of Discrete-Time Sliding Mode Control

The design of DTSM control for the system (5.5) is discussed here. The discrete-time dynamics of sliding variable can be written as

$$s(k+1) = c^\top z(k+1)$$
$$= c^\top \Phi_\tau z(k) + c^\top \Gamma_\tau u(k) + c^\top \Gamma_\tau d(k). \tag{5.11}$$

We define $\tilde{d} = c^\top \Gamma_\tau d$. Then, the DTSM control can be designed using Bartoszewicz's reaching law that brings the system trajectory to the vicinity of sliding manifold. It is given below as

$$u(k) = -\left(c^\top \Gamma_\tau\right)^{-1}\left(c^\top \Phi_\tau z(k) - s_d(k+1)\right). \tag{5.12}$$

On substituting (5.12) in (5.11) for control $u(k)$, the closed loop sliding dynamics can be written as

$$s(k+1) = \tilde{d}(k) + s_d(k+1). \tag{5.13}$$

This is same as the relation (5.16). When $s_d(k)$ decreases in each discrete steps, the sliding variable also decreases due to (5.13). Once the discrete-time steps, k, equals $k^\star$, the sliding dynamics (5.13) becomes

$$s(k+1) = \tilde{d}(k), \qquad \text{for all} \quad k \geq k^\star.$$

From the above relation, the QSM band can be obtained as

$$\left\{ z \in \mathbb{R}^n : |s| = \left|c^\top z\right| \leq \tilde{d}_0 \right\}. \tag{5.14}$$

One of the important observations is that the exact sliding mode is not possible in the case of discrete-time SMC control. Here, the trajectories remain bounded in the vicinity of sliding manifold which depend on the disturbance bound. This is achieved by updating the control signal at every periodic time interval. But, this may not be economical from the practical point of view due to more energy expenditure and more computational burden. So, the objective is to show the QSM band is designed to be minimum with less control computation using event-triggering strategy.

5.3 Discrete Event-Triggered Sliding Mode Control: State Feedback Approach

In sampled-data system, the control signal is updated periodically to the plant. There is another control implementation strategy, namely event-triggering, where the control may be implemented without losing the system stability. Since the system is already in discrete-time domain, the event-triggering rule can be evaluated only at discrete-time instants. So, the triggering instant is always a multiple of sampling interval (τ in this case). Here, it is assumed that all the states are available for the control design.

We denote $\{k\tau\}_{k \in \mathbb{Z}_{\geq 0}} = \{k\}_{k \in \mathbb{Z}_{\geq 0}}$ for a given constant time interval $\tau > 0$. Let $\{k_i\}_{i \in \mathbb{Z}_{\geq 0}}$ be the sequence of triggering instants for updating control. So, the DTSM control law (5.12) corresponding to the triggering sequence $\{k_i\}_{i \in \mathbb{Z}_{\geq 0}}$ can be written as

$$u(k) = - \left(c^\top \Gamma_\tau\right)^{-1} \left(c^\top \Phi_\tau z(k_i) - s_d(k_i + 1)\right) \tag{5.15}$$

for all $k \in [k_i, k_{i+1})$. If the control is updated at every τ interval, the time instant k_{i+1} equals $k + 1$. However, this is not the case with event-triggering scheme. The priori known function $s_d(k)$ in (5.15) changes at every time instant k_i. So, in the time interval $[k_i, k_{i+1})$, the variable $s_d(k)$ decreases by the same amount in the triggering interval as it would have decreased in every k instants in periodic implementation. This ensures that the convergence to sliding manifold is not affected due to event-triggering scheme. We define $e(k) = z(k_i) - z(k)$ as the error introduced in the system due to implementation of control (5.15) at the sequence of time instants $\{k_i\}_{i \in \mathbb{Z}_{\geq 0}}$.

It is very natural that when the control is implemented at the triggering instant, the system trajectory remains bounded in the vicinity of the sliding manifold. But, the QSM band in this case need not be the same as it is obtained in the periodic implementation. This is because in the event-triggered implementation the control is updated at aperiodic instants which is greater than or equal to the periodic sampling period. As a result, the sliding mode band obtained in event-triggered strategy is larger than that of the periodic implementation. It may be noted that in the discrete event-triggering strategy the frequent control computation is reduced by restricting the trajectories within a band slightly larger than QSM band.

Definition 5.2 *(Practical Quasi-Sliding Mode)* The system (5.2) is said to be in practical QSM if the system is in QSM with band size greater than the QSM band (ε). That means for some $\varepsilon_1 > \varepsilon > 0$ there exists $\bar{k} > 0$ such that the sliding trajectory $s(k) = c^\top z(k)$ satisfies

$$|s(k)| \leq \varepsilon_1$$

for all $k \geq \bar{k}$. The constant $\varepsilon_1 > 0$ is the size of practical QSMB.

From the above, it implies that the band size of practical QSM is always greater than or equals that of QSM. In order to achieve this practical QSM, the control law needs to be designed. Since we are using Bartoszewicz's reaching law for designing the control law, first the Bartoszewicz's reaching law is stated for event-triggering scheme. We call this as the event-triggered Bartoszewicz's reaching law.

5.3.1 Event-Triggered Bartoszewicz's Reaching Law

The original Bartoszewicz's reaching law is modified to design the event-triggered DTSM for the discrete-time system. This is done by introducing a constant $\alpha > 0$ in the reaching law which is given below as

$$s(k+1) = \tilde{d}(k) + \alpha + s_d(k+1) \tag{5.16}$$

where $\tilde{d}(k)$ and $s_d(k)$ are defined as earlier, and $\alpha > 0$. Then the followings hold:

- If $|s(0)| > 2(\tilde{d}_0 + \alpha)$, then

$$s_d(0) = s(0)$$
$$s_d(k) \cdot s(0) \geq 0, \quad \text{for any } k \geq 0$$
$$s_d(k) = 0, \quad \text{for any } k \geq k^\star$$
$$|s_d(k+1)| < |s_d(k)| - 2(\tilde{d}_0 + \alpha), \quad \text{for any } k < k^\star.$$

- Otherwise, i.e., if $|s(0)| \leq 2(\tilde{d}_0 + \alpha)$, then $s_d(k) = 0$ for any $k \geq 0$.

This reaching law ensures that in each time step the sliding variable decreases by a magnitude greater than $2(\tilde{d}_0 + \alpha)$ instead of $2\tilde{d}_0$. Thus, the function $s_d(k)$ in the case of event-triggered reaching law is also modified and is given as

$$s_d(k) = \frac{k^\star - k}{k^\star} s(0), \qquad k^\star < \frac{|s(0)|}{2(\tilde{d}_0 + \alpha)}.$$

Here, the constant α is a design parameter and is used for designing the triggering rule which will be discussed later. Since the control signal is not updated periodically in regular intervals of time, the state trajectory may increase in each successive discrete-time steps until the control is updated. But, the trajectory is always remained bounded. This is established in the following Theorem.

Theorem 5.1 *Consider the system (5.5) and the control law (5.15). Let $\{k_i\}_{i \in \mathbb{Z}_{\geq 0}}$ be the sequence of time instants at which control is updated. Let $\alpha > 0$ be given. Then the system is in practical quasi-sliding mode if*

$$\left| c^\top \Phi_\tau e(k) \right| < \alpha \tag{5.17}$$

for all $k \in \mathbb{Z}_{\geq 0}$.

Proof First, let $k \in [k_i, k_{i+1})$ such that $k_i < k^\star$. Then, the sliding dynamics can be obtained with the control (5.15),

$$
\begin{aligned}
s(k+1) &= c^\top z(k+1) \\
&= c^\top \Phi_\tau z(k) + c^\top \Gamma_\tau u(k) + c^\top \Gamma_\tau d(k) \\
&= -c^\top \Phi_\tau e(k) + \tilde{d}(k) + s_d(k_i + 1).
\end{aligned} \tag{5.18}
$$

This can be further reduced to

$$
\begin{aligned}
s(k+1) &\leq \left| c^\top \Phi_\tau e(k) \right| + \left| \tilde{d}(k) \right| + s_d(k_i + 1) \\
&< \alpha + \tilde{d}_0 + s_d(k_i + 1).
\end{aligned} \tag{5.19}
$$

To see how $s(k)$ converges to the practical QSMB, consider the evolution of $s_d(k)$. Note that the control is not updated until the time instant k_{i+1}, so $s_d(k)$ changes at k_{i+1}. By definition of reaching law, $s_d(k)$ decreases by a magnitude at least equal to $2(\tilde{d}_0 + \alpha)$ in every (periodic) discrete-time step. Due to this, the function $s_d(k)$ decreases by $2(k_{i+1} - k_i)(\tilde{d}_0 + \alpha)$ in the time interval $[k_i, k_{i+1})$. This continues until the practical QSM band is reached. When $k_i \geq k^\star$, then for all $k^\star \leq k \in [k_i, k_{i+1})$, we have $s_d(k) = 0$. Thus, the sliding dynamics (5.19) can be written as

$$s(k) < \alpha + \tilde{d}_0, \qquad \forall k \geq k^\star.$$

This shows that sliding trajectory remains bounded by $\alpha + \tilde{d}_0$ for all $k \geq k^\star$. This is defined as practical QSM band and is given as

$$\Omega = \left\{ z \in \mathbb{R}^n : |s| = \left| c^\top z \right| < \varepsilon_1 \right\} \tag{5.20}$$

where $\varepsilon_1 := \alpha + \tilde{d}_0$. The proof is completed. $\qquad\qquad\square$

The system trajectory moves towards the origin while remaining within the practical QSM band. This can be shown using sliding mode dynamics when the system is in practical QSM. The sliding mode dynamics can be given as

$$z_1(k+1) = \Phi_{\tau_{11}} z_1(k) + \Phi_{\tau_{12}} z_2(k) \tag{5.21}$$

$$z_2(k) = -c_1^\top z_1(k) + s(k). \tag{5.22}$$

So, the sliding dynamics can be written as

$$z_1(k+1) = \left(\Phi_{\tau_{11}} - \Phi_{\tau_{12}} c_1^\top \right) z_1(k) + \Phi_{\tau_{12}} s(k). \tag{5.23}$$

Since the matrix $\Phi_{\tau_{11}} - \Phi_{\tau_{12}} c_1^\top$ Schur stable by design and the bounded sliding trajectory $s(k)$ due to Theorem 5.1, the system trajectory also remains bounded.

5.3.2 Event-Triggering Rule

In this section, the triggering scheme is presented for implementing the DTSM control using event-triggering strategy. We see that if the system is in practical QSM, then the system stability is ensured. It is seen from Theorem 5.1 that the relation (5.17) is essential for stability, so it may be used to generate the triggering rule for the system (5.5). Indeed, in practice a more stronger condition is desired rather than actual relation to provide margin for unavoidable delay. So, the relation $|c^\top \Phi_\tau e| < \sigma\alpha$ is considered for some $\sigma \in (0, 1)$ in order to compensate for unavoidable delays/computations. Thus, the triggering scheme can now be given as

$$k_{i+1} = \inf\left\{k > k_i : \|c\|\|\Phi_\tau\|\|e(k)\| \geq \sigma\alpha\right\}. \tag{5.24}$$

The triggering rule (5.24) always generated whenever $|c^\top \Phi_\tau e| < \sigma\alpha$ is violated. That means, this relation is always satisfied for all steps. So, the results of Theorem 5.1 also holds since (5.17) is also true.

5.3.3 Simulation Results

Consider the dynamical system

$$\dot{x} = \begin{bmatrix} 0 & 1 \\ 4 & 5 \end{bmatrix} x + \begin{bmatrix} 0 \\ 1 \end{bmatrix} (u + d)$$
$$y = \begin{bmatrix} 1 & 0 \end{bmatrix} x.$$

The sampling time period τ is selected as 0.1 for ZOH discretization model of the plant. Clearly, the pair $(\overline{\Phi}_\tau, \overline{\Gamma}_\tau)$ is controllable. The transformation matrix T is chosen such that it transforms the system $\{\overline{\Phi}_\tau, \overline{\Gamma}_\tau, C\}$ into regular form $\{\Phi_\tau, \Gamma_\tau, C_\tau\}$ as

$$T = \begin{bmatrix} 1 & -0.0457 \\ 0 & 1 \end{bmatrix}.$$

The sliding surface vector is chosen as $c = \begin{bmatrix} 0.9 & 1 \end{bmatrix}^\top$ such that the sliding mode dynamics renders stability. The event parameters are chosen as $\alpha = 0.1$ and $\sigma = 0.8$. The disturbance is taken as $d = 0.01 \sin t$ for simulation. The initial condition

Fig. 5.1 Performance of discrete event-triggered SMC for LTI systems

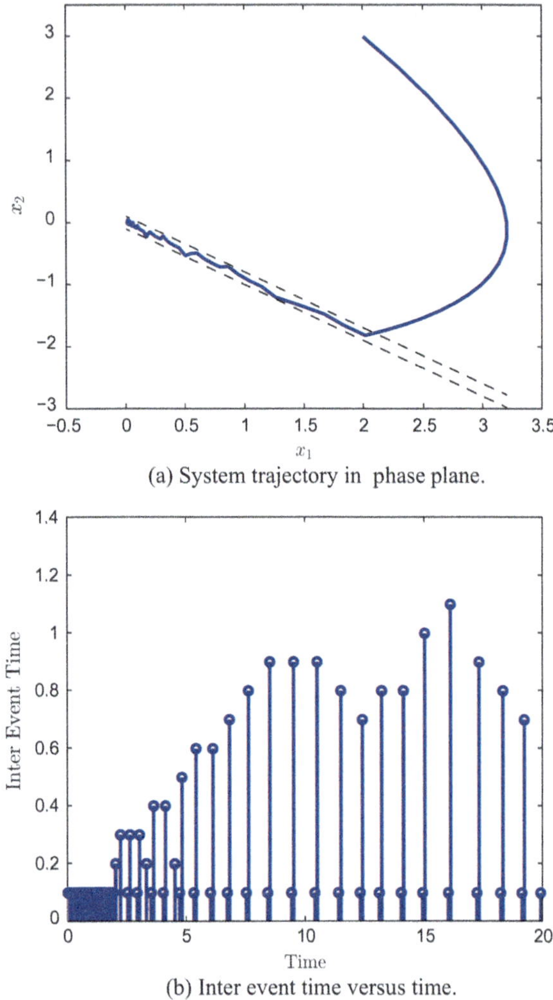

(a) System trajectory in phase plane.

(b) Inter event time versus time.

is chosen as $z_0 = \begin{bmatrix} 1 & 2 & 3 \end{bmatrix}^\top$. The value of $k^\star$ is selected as 20 which is less than $\frac{|s(0)|}{2(\tilde{d}_0+\alpha)} = 23$.

The performance of the discrete event-triggered SMC is shown in Figs. 5.1 and 5.2. The system trajectories remain bounded within the band $\alpha + \tilde{d}_0 = 0.103$ as shown in Fig. 5.1a. Thus, practical QSM is ensured in the system after 20 discrete-time steps. The variation of sampling interval is plotted in Fig. 5.1b. It can be seen that there always exists a minimum lower bound of 0.1 while the maximum of it is as high as 1.1. The control signal is also shown in Fig. 5.2. This shows that the control remains constants until event is triggered to update the control signal.

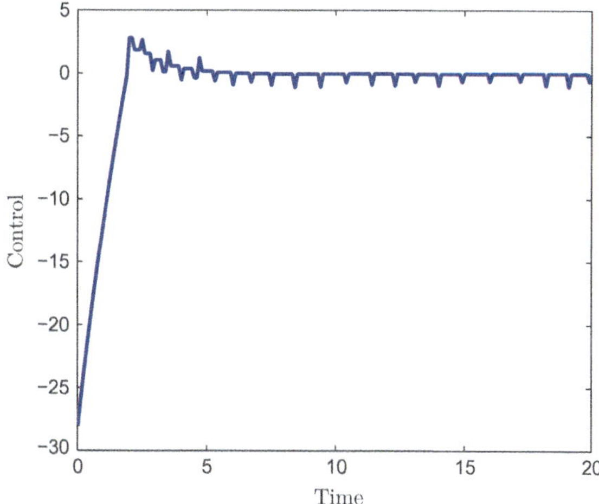

Fig. 5.2 Discrete event-triggered SMC signal for LTI systems

5.4 Discrete Event-Triggered Sliding Mode Control: Output Feedback Approach

In this section, the discrete event-triggered SMC is presented with output feedback. Here, it is assumed that only output measurements are available for the control design. It is possible that an observer may be designed for the control design. But, in this chapter a static output feedback-based control law is designed instead of using an observer. Multirate output feedback (MROF) technique is employed to obtain the state measurement in each sampling technique to facilitate the control design. Out of the many MROF technique, in this case fast output sampling (FOS) approach is used where the outputs are sampled at a faster rate than the control input is applied to the plant. It is seen that in this strategy, there always exists a output feedback gain such that the closed loop system is stable. First, we briefly present the FOS-based MROF technique.

5.4.1 Multirate Output Feedback Technique

Consider the dynamical system (5.1). Let the output of this system is sampled at every Δ interval and input is applied to the plant at every τ interval. To analyse the stability of the system, consider ZOH discretization of the system given by (5.2). Since the system (5.1) evolves in continuous-time, the dynamics of the system can be evaluated at any sampling interval. Let the discrete-time system, sampled at every Δ interval using ZOH discretization of (5.1), is given as

$$x(k+1) = \overline{\Phi}_\Delta x(k) + \overline{\Gamma}_\Delta u(k) + \overline{\Gamma}_\Delta d(k) \tag{5.25}$$

$$y(k) = Cx(k) \tag{5.26}$$

where $\overline{\Phi}_\Delta$ and $\overline{\Gamma}_\Delta$ are given by (5.4) with $\tau = \Delta$. Note that in the ZOH discretization it is assumed that the disturbance $d(t)$ remains almost constant over τ interval. So, this assumption is also valid for the dynamics evolving in every Δ interval. Here, $x(k)$ means $x(k\Delta)$, so the state evolves in every Δ interval. The same is true for all other variables. In the below, the following assumption is needed to hold to design MROF-based control.

Assumption 5.2 The pairs $(\overline{\Phi}_\Delta, \overline{\Gamma}_\Delta)$ and $(\overline{\Phi}_\Delta, C)$ are controllable and observable, respectively.

Here, time period Δ is chosen as $\Delta = \tau/N$ where N is greater than or equal to the observability index of the system (5.25) and (5.26). Before proceeding to design the MROF-based control, the system (5.25) and (5.26) must be first transformed into the transformed domain of τ-dynamics given by (5.5). Now using the transformation T, we obtain

$$z(k+1) = \Phi_\Delta z(k) + \Gamma_\Delta u(k) + \Gamma_\Delta d(k) \tag{5.27}$$

$$y(k) = C_\Delta z(k) \tag{5.28}$$

where $\Phi_\Delta = T\overline{\Phi}_\Delta T^{-1}$, $\Gamma_\Delta = T\overline{\Gamma}_\Delta$ and $C_\Delta = CT^{-1}$. Let the output is sampled at Δ intervals of time. During every τ interval, there is N output measurements. These N output measurements are enough to obtain the state information. Note that in Δ-dynamics $u(k)$ and $d(k)$ does not change by assumption. First, writing the Δ-dynamics in $[k\tau, (k+1)\tau]$ interval as

$$z(k\tau + \Delta) = \Phi_\Delta z(k\tau) + \Gamma_\Delta u(k\tau) + \Gamma_\Delta d(k\tau)$$

$$z(k\tau + 2\Delta) = \Phi_\Delta z(k\tau + \Delta) + \Gamma_\Delta u(k\tau) + \Gamma_\Delta d(k\tau)$$

$$= \Phi_\Delta^2 z(k\tau) + \sum_{i=0}^{1} \Phi_\Delta^i \Gamma_\Delta u(k\tau) + \sum_{i=0}^{1} \Phi_\Delta^i \Gamma_\Delta d(k\tau)$$

$$z(k\tau + 3\Delta) = \Phi_\Delta z(k\tau + 2\Delta) + \Gamma_\Delta u(k\tau) + \Gamma_\Delta d(k\tau)$$

$$= \Phi_\Delta^3 z(k\tau) + \sum_{i=0}^{2} \Phi_\Delta^i \Gamma_\Delta u(k\tau) + \sum_{i=0}^{2} \Phi_\Delta^i \Gamma_\Delta d(k\tau)$$

$$\vdots$$

$$z(k\tau + N\Delta) = \Phi_\Delta^N z(k\tau) + \sum_{i=0}^{N-1} \Phi_\Delta^i \Gamma_\Delta u(k\tau) + \sum_{i=0}^{N-1} \Phi_\Delta^i \Gamma_\Delta d(k\tau).$$

From the last relation, we obtain τ-dynamics and comparing with the dynamics (5.5), one obtains

$$\Phi_\tau = \Phi_\Delta^N \quad \text{and} \quad \Gamma_\tau = \sum_{i=0}^{N-1} \Phi_\Delta^i \Gamma_\Delta.$$

Now in a similar way, writing the output measurements during the interval $[k\tau, (k+1)\tau)$ as

$$y(k\tau) = C_\Delta z(k\tau)$$
$$y(k\tau + \Delta) = C_\Delta \Phi_\Delta z(k\tau) + C_\Delta \Gamma_\Delta u(k\tau) + C_\Delta \Gamma_\Delta d(k\tau)$$
$$y(k\tau + 2\Delta) = C_\Delta \Phi_\Delta^2 z(k\tau) + \sum_{i=0}^{1} C_\Delta \Phi_\Delta^i \Gamma_\Delta u(k\tau) + \sum_{i=0}^{1} C_\Delta \Phi_\Delta^i \Gamma_\Delta d(k\tau)$$
$$\vdots$$
$$y(k\tau + (N-1)\Delta) = C_\Delta \Phi_\Delta^{N-1} z(k\tau) + \sum_{i=0}^{N-2} C_\Delta \Phi_\Delta^i \Gamma_\Delta u(k\tau) + \sum_{i=0}^{N-2} C_\Delta \Phi_\Delta^i \Gamma_\Delta d(k\tau).$$

These all output measurements can be obtained from the sensor measurements. Now, stacking all these outputs, we write

$$y_{k+1} = \mathscr{C}_0 z(k) + \mathscr{D}_0 u(k) + \mathscr{D}_0 d(k) \tag{5.29}$$

where

$$\mathscr{C}_0 = \begin{bmatrix} C_\Delta \\ C_\Delta \Phi_\Delta \\ \vdots \\ C_\Delta \Phi_\Delta^{N-1} \end{bmatrix}, \quad \mathscr{D}_0 = \begin{bmatrix} 0 \\ C_\Delta \Gamma_\Delta \\ \vdots \\ C_\Delta \sum_{i=0}^{N-2} \Phi_\Delta^i \Gamma_\Delta \end{bmatrix}, \quad \text{and} \quad y_{k+1} := \begin{bmatrix} y(k\tau) \\ y(k\tau + \Delta) \\ \vdots \\ y((k+1)\tau - \Delta) \end{bmatrix}.$$

Thus, the system dynamics at every τ-th time instant can be obtained as

$$z(k+1) = \Phi_\tau z(k) + \Gamma_\tau u(k) + \Gamma_\tau d(k) \tag{5.30}$$
$$y_{k+1} = \mathscr{C}_0 z(k) + \mathscr{D}_0 u(k) + \mathscr{D}_0 d(k). \tag{5.31}$$

From (5.31), we write $z(k) = (\mathscr{C}_0^\top \mathscr{C}_0)^{-1} \mathscr{C}_0^\top (y_{k+1} - \mathscr{D}_0 u(k) - \mathscr{D}_0 d(k))$. The matrix $\mathscr{C}_0$ is full rank by assumption of observability. Substituting this in (5.30),

$$z(k+1) = L_y y_{k+1} + L_u u(k) + L_u d(k) \tag{5.32}$$

where

$$L_y = \Phi_\tau (\mathscr{C}_0^\top \mathscr{C}_0)^{-1} \mathscr{C}_0^\top$$
$$L_u = \Gamma_\tau - \Phi_\tau (\mathscr{C}_0^\top \mathscr{C}_0)^{-1} \mathscr{C}_0^\top \mathscr{D}_0.$$

Note that if there is no disturbance, i.e., $d(k) = 0$ for all $k \in \mathbb{Z}_{\geq 0}$, then the state can be completely recovered from the following relation

$$z(k) = L_y y_k + L_u u(k - 1).$$

This relation gives the actual state information from the past sampled outputs and control input. It may be noted that for $k = 0$ the state $z(0)$ cannot be obtained from the above relation as the past output and the past control input are not available. So, for this case, any arbitrary $z(0)$ may be given initially to compute the output feedback control, and then use the above relation for subsequent control computation. However, in this technique some error is introduced in the system due to incorrect feedback at the initial time instant and it may propagate for subsequent time instants. But, in case of the linear feedback law, it is shown that this error becomes zero in the next immediate step for the system provided there is no uncertainty. For DTSM, the stability of the system is also assured as discussed below.

The relation (5.32) gives the relation between output measurements, past control input and disturbance and state. Since the disturbance is present in the system, the relation (5.32) may not be used directly to estimate the state. Nevertheless the following relation can be used instead of (5.32) to ensure the stability

$$\bar{z}(k) = L_y y_k + L_u u(k - 1). \tag{5.33}$$

From this, we see that $z(k) = \bar{z}(k) + L_u d(k - 1)$. In the next, the design of FOS-based DTSM control is discussed.

5.4.2 Multirate-Based Event-Triggered Discrete-Time Sliding Mode

In this section, the DTSM based on MROF is briefly presented first, and then, event-triggering-based design of SMC is discussed.

The basic difference between state feedback and MROF-based DTSM is that the relation (5.33) is used in place of $z(k)$ in the control law (5.12). Since the uncertain term appears in actual state expression (5.32), the Bartoszewicz's reaching law contains one more uncertain term as given below

$$s(k + 1) = \tilde{d}(k) + f(k - 1) + s_d(k + 1) \tag{5.34}$$

where $f(k) = c^\top \Phi_\tau L_u d(k)$ is the uncertainty with the known bound $\sup_{k \geq 0} |f(k)| \leq f_0$. The FOS-based DTSM control obtained from the reaching law (5.34) is given as

$$u(k) = -\left(c^\top \Gamma_\tau\right)^{-1}\left(c^\top \Phi_\tau \bar{z}(k) - s_d(k+1)\right) \tag{5.35}$$

where $\bar{z}(k)$ is obtained from the relation (5.32). Note that there is no uncertainty associated with the control law, so this control can be implemented easily. Another important difference between the reaching law (5.16) and the reaching law (5.34) is that the convergence time now to the QSM band is given by

$$k^\star < \frac{|s(0)|}{2\tilde{f}_0}$$

instead of $k^\star < \frac{|s(0)|}{2\tilde{d}_0}$. Here, the constant $\tilde{f}_0$ satisfies $\sup_{k \geq 0} |\tilde{d}(k) + f(k-1)| \leq \tilde{f}_0 := \tilde{d}_0 + f_0$. This is readily obtained by invoking the conditions of Bartoszewicz's reaching law. As a result of this, the size of QSM band is increased due to the uncertain term $f(k)$ in (5.34).

The main objective is to develop event-triggering DTSM using MROF technique. Consider the DTSM control (5.35). Now, it is very natural to use the relation (5.33) in the place of $\bar{z}(k_i)$. At every τ interval of time, the state is computed using FOS technique and then event is evaluated at these time instants for possible triggering instant. If the event is detected then only the control is updated at the plant, otherwise the output is sampled for next τ interval for evaluating the event. The event-triggered DTSM control using MROF can be written using the relation (5.33) as

$$u(k) = -\left(c^\top \Gamma_\tau\right)^{-1}\left(c^\top \Phi_\tau \bar{z}(k_i) - s_d(k_i + 1)\right) \tag{5.36}$$

for all $k \in [k_i, k_{i+1})$. The sliding dynamics with the control law (5.36) becomes

$$s(k+1) = -c^\top \Phi_\tau \bar{e}(k) + \tilde{d}(k) + f(k-1) + s_d(k_i + 1) \tag{5.37}$$

where $\bar{e}(k) = \bar{z}(k_i) - \bar{z}(k)$. Define $\tilde{f}(k) = \tilde{d}(k) + f(k-1)$ and recall that $\sup_{k \geq 0} |\tilde{f}(k)| \leq \tilde{f}_0$. Similar to (5.34), the convergence to QSM band again will constrained to

$$k^\star < \frac{|s(0)|}{2(\tilde{f}_0 + \alpha)}$$

for some $\alpha > 0$ satisfying $|c^\top \Phi_\tau \bar{e}(k)| < \alpha$. When $k^\star$ is selected as per this, the sliding trajectory reaches the QSM band in $k^\star$-th step. So, for all $k \geq k^\star$, the sliding dynamics becomes $s(k+1) = -c^\top \Phi_\tau \bar{e}(k) + \tilde{f}(k)$. This is summarized in the following Theorem.

Theorem 5.2 *Consider the system (5.5) and the control law (5.36). Let $\{k_i\}_{i\in\mathbb{Z}_{\geq 0}}$ be the sequence of time instant at which control is updated. Let $\alpha > 0$ be given. Then the system achieves practical QSM if*

$$\left| c^\top \Phi_\tau \bar{e}(k) \right| < \alpha \tag{5.38}$$

for all $k \in \mathbb{Z}_{\geq 0}$.

Proof Proof follows the similar lines as in Theorem 5.1. $\qquad\square$

Remark 5.1 Another most important observation in MROF-based event-triggered DTSM is that the error $\bar{e}(k)$ is evaluated rather than $e(k)$ as it is done in state feedback.

So, the event-triggering rule is also changed due to the relation (5.38). This relation is essential to ensure practical QSM in the system. The event-triggering rule is given below as

$$k_{i+1} = \inf \{k > k_i : \|c\| \|\Phi_\tau\| \|\bar{e}(k)\| \geq \sigma\alpha\}$$

for some $\sigma \in (0, 1)$. Once the system is in practical QSM, the stability of sliding mode dynamics can also be shown in the similar manner as in the state feedback case.

5.4.3 Simulation Results

Consider a discrete-time dynamical system

$$\dot{x} = \begin{bmatrix} 0 & 1 \\ 4 & 5 \end{bmatrix} x + \begin{bmatrix} 0 \\ 1 \end{bmatrix} (u + d)$$
$$y = \begin{bmatrix} 1 & 0 \end{bmatrix} x.$$

The values of τ and Δ are selected as 0.1 and 0.05 with $N = 2$, respectively. First, the discrete-time system is transformed into regular form using the nonsingular transformation matrix T as obtained in state feedback case. All the parameters remain same as that of state feedback case. Reaching time to sliding manifold is given as $k^\star = 30$. The event parameters are $\alpha = 0.1$ and $\sigma = 0.85$. The simulation is run with the initial condition $z_0 = [6\ 4]^\top$.

The performance of the system with FOS feedback event-triggered SMC is shown in Figs. 5.3 and 5.4. The system trajectory reaches the sliding manifold in finite number of steps, in this case $k^\star = 30$ steps. When it reaches the manifold, it remains bounded within the practical QSM band. Similarly, the corresponding inter-event times are shown in Fig. 5.3b. It shows that the control is updated almost more than the 0.1 unit. It remains constant until the next triggering constant.

Fig. 5.3 Performance of discrete event-triggered SMC for LTI systems

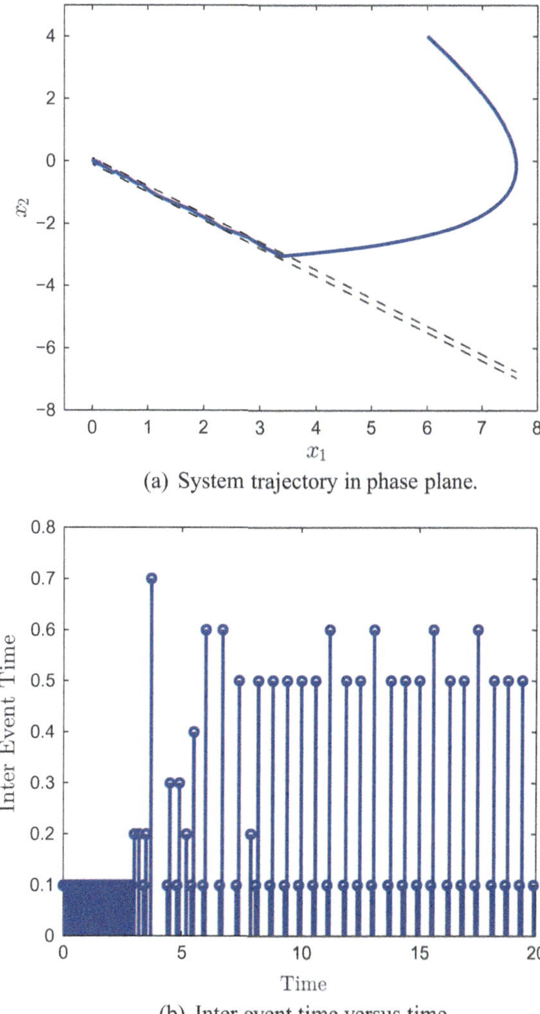

(a) System trajectory in phase plane.

(b) Inter event time versus time.

5.5 Summary

An event-triggering scheme for DTSM control is proposed in this chapter. In this technique, the triggering rule is checked only at discrete instants of time and control is updated whenever it is violated. However, the band size is increased due to this event-triggering scheme but remains bounded. The major advantage in this technique is that there is no Zeno execution of triggering sequences. Further, output feedback-

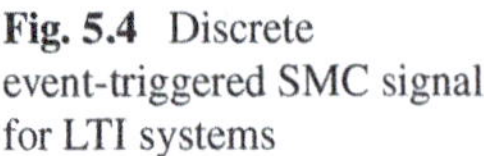

Fig. 5.4 Discrete event-triggered SMC signal for LTI systems

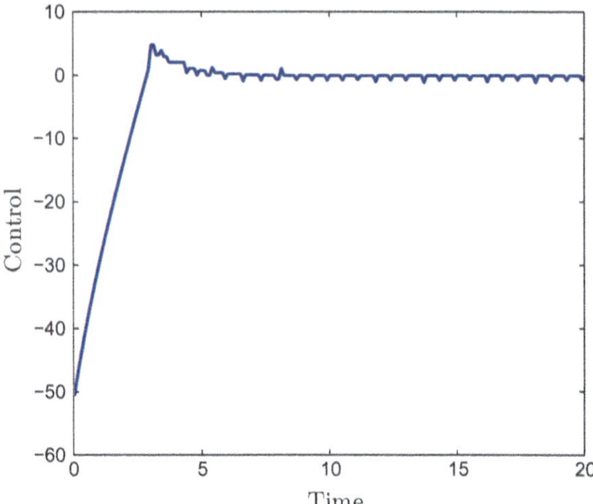

based discrete event-triggered SMC is analyzed using FOS. The outputs are sampled periodically and, the event is checked for possible generation of triggering instant. In case there is no triggering, again the outputs are sampled until next discrete instant. This establishes a feasible output feedback-based discrete event-triggered SMC.

5.6   Notes and References

The design of SMC and DTSM control is discussed in detail in [1–12]. The output feedback-based design of DTSM using multirate technique is given in [13–15]. The continuous event-triggering-based design of SMC for LTI systems can be found in [20–23]. The preliminaries on event-triggered control strategy are presented in [16–19]. The event-triggering strategy using periodic measurements is discussed in [24–26] and references therein. In [27–29], the design of discrete event-triggered SMC with periodic measurements is discussed in more detail.

References

1. V.I. Utkin, Variable structure systems with sliding modes. IEEE Trans. Autom. Control **22**(2), 212–222 (1977)
2. V.I. Utkin, *Sliding Modes and Their Applications in Variable Structure Systems*, Translated from the Russian by A. Parnakh (MIR Publishers, Moscow, 1978)
3. D. Draženović, The invariance conditions in variable structure systems. Automatica **5**(3), 287–295 (1969)
4. C. Edwards, S.K. Spurgeon, *Sliding Mode Control: Theory and Applications* (CRC Press, Taylor and Francis Group, 1998)

5. S.Z. Sarpturk, I. Istefanopulus, O. Kaynak, On the stability of discrete-time sliding mode control systems. IEEE Trans. Autom. Control **32**(10), 930–932 (1987)
6. K. Furuta, Sliding mode control of a discrete system. Syst. Control Lett. **14**(2), 145–152 (1990)
7. W. Gao, Y. Wang, A. Homaifa, Discrete-time variable structure control systems. IEEE Trans. Industr. Electron. **42**(2), 117–122 (1995)
8. G. Bartolini, A. Ferrara, V.I. Utkin, Adaptive sliding mode control in discrete-time systems. Automatica **31**(5), 769–773 (1995)
9. W.-C. Su, S.V. Drakunov, Ü. Özgüner, An $O(T^2)$ boundary layer in sliding mode for sampled-data systems. IEEE Trans. Autom. Control **45**(3), 482–485 (2000)
10. A. Bartoszewicz, Discrete-time quasi-sliding-mode control strategies. IEEE Trans. Industr. Electron. **45**(4), 633–637 (1998)
11. A. Bartoszewicz, P. Lesniewski, New switching and nonswitching type reaching laws for SMC of discrete time systems. IEEE Trans. Control Syst. Technol. **24**(2), 670–677 (2016)
12. S. Chakrabarty, B. Bandyopadhyay, A generalized reaching law for discrete time sliding mode control. Automatica **52**, 83–86 (2015)
13. M.C. Saaj, B. Bandyopadhyay, H. Unbehauen, A new algorithm for discrete-time sliding-mode control using fast output sampling feedback. IEEE Trans. Industr. Electron. **49**(3), 518–523 (2002)
14. S. Janardhanan, B. Bandyopadhyay, Output feedback sliding-mode control for uncertain systems using fast output sampling technique. IEEE Trans. Industr. Electron. **53**(5), 1677–1682 (2006)
15. B. Bandyopadhyay, S. Janardhanan, Discrete-Time Sliding Mode Control- A Multirate Output Feedback Approach. Lecture Notes in Control and Information Sciences, vol. 323 (Springer, Berlin, 2005), p. 150
16. K.-E. Årźen, A simple event-based PID controller, in *Proceedings of 14th IFAC World Congress*, Beijing, China (1999), pp. 423–428
17. K.J. Åström, B. Bernhardsson, Comparison of Riemann and Lebesgue sampling for first order stochastic systems, in *Proceedings of 41st IEEE Conference on Decision and Control*, Las Vegas, USA (1992), pp. 2011–2016
18. P. Tabuada, Event-triggered real-time scheduling of stabilizing control tasks. IEEE Trans. Autom. Control **52**(9), 1680–1685 (2007)
19. W.P.M.H. Heemels, K.H. Johansson, P. Tabuada, An introduction to event-triggered and self-triggered control, in *Proceedings of 51st IEEE Conference of Decision and Control*, Hawai, USA (2010), pp. 3270–3285
20. A.K. Behera, B. Bandyopadhyay, Event based robust stabilization of linear systems, in *Proceedings of 40th Annual Conference of the IEEE Industrial Electronics Society*, Dallas, USA (2014), pp. 133–138
21. A.K. Behera, B. Bandyopadhyay, Self-triggering-based sliding-mode control for linear systems. IET Control Theory Appl. **9**(17), 2541–2547 (2015)
22. A.K. Behera, B. Bandyopadhyay, Event-triggered sliding mode control for a class of nonlinear systems. Int. J. Control **89**(9), 1916–1931 (2016)
23. A.K. Behera, B. Bandyopadhyay, Robust sliding mode control: an event-triggering approach. IEEE Trans. Circuits Syst. II: Express Briefs **64**(2), 146–150 (2017)
24. W.P.M.H. Heemels, M.C.F. Donkers, Model-based periodic event-triggered control for linear systems. Automatica **49**(3), 698–711 (2013)
25. W.P.M.H. Heemels, M.C.F. Donkers, A.R. Teel, Periodic event-triggered control for linear systems. IEEE Trans. Autom. Control **58**(4), 847–861 (2013)
26. R. Postoyan, A. Anta, W.P.M.H. Heemels, P. Tabuada, D. Nešić, Periodic event-triggered control for nonlinear systems, in *Proceedings of 52nd IEEE Conference on Decision and Control*, Florence, Italy (2013), pp. 7397–7402
27. A.K. Behera, B. Bandyopadhyay, J. Reger, Discrete event-triggered sliding mode control with fast output sampling feedback, in *Proceedings of 14th IEEE International Workshop on Variable Structure Systems*, Nanjing, China (2016), pp. 148–153

28. K. Kumari, B. Bandyopadhyay, A.K. Behera, J. Reger, Event-triggered sliding mode control for delta opeartor systems, in *Proceedings of 42nd Annual Conference of IEEE Industrial Electronics Society*, Florence, Italy (2016), pp. 148–153
29. A.K. Behera, B. Bandyopadhyay, Discrete event-triggered sliding mode control, in *Advances in Variable Structure Systems and Sliding Mode Control: Theory and Applications*, ed. by S. Li, X. Yu, L. Fridman, Z. Man, X. Wang. Studies in Systems, Decision and Control, Springer International Publishing, Cham, Switzerland, vol. 115 (2018), pp. 289–304

Chapter 6
Event-Triggered Sliding Mode Control with Quantized State Measurements

Control with limited resources has been studied extensively since the last decade to optimize the system performance in terms of economic and practical constraints [8, 9]. In many practical applications, the state measurements available at the controller end are quantized values and the control is designed based on these quantized measurements. The stability of the systems with quantized information has been studied in [8–10]. It is well known that with quantized feedback asymptotic stabilization of systems cannot be guaranteed. But, some strategies have been proposed by introducing dynamical quantizers to achieve the system performance similar to the case of unquantized with linear state feedback in [8] and with SMC in [11, 12].

In this work, we analyze the event-triggering-based implementation of SMC when quantized measurements are available for updating the control signal. Throughout this chapter, we assume that the quantizer is a static one with some fixed saturation level. Our design approach assumes that the event generator is located at the sensor end, and thus, the actual state measurements are available for event-triggering ensuring system stability. The control task is executed only when a triggering instant is generated. Moreover, by suitable design of event parameter, it can be ensured that new quantized measurements are transmitted every time when the event is detected. First, the stability of SMC with quantized state measurements is shown. Then, the stability of the event-triggered SMC using quantized measurements is presented.

6.1 System Description

Consider the SISO LTI system as given by (1.11),

$$\dot{x} = Ax + B(u + d), \qquad x_0 = x(t_0)$$

© Springer International Publishing AG, part of Springer Nature 2018

B. Bandyopadhyay and A. K. Behera, *Event-Triggered Sliding Mode Control*, Studies in Systems, Decision and Control 139, https://doi.org/10.1007/978-3-319-74219-9_6

where $x \in \mathbb{R}^n$ and $u \in \mathbb{R}$ are the states and control input to the system, respectively. The Assumption 1.1 on disturbance $d(t)$ also holds. In this chapter, we focus on the stability of event-triggered SMC when the quantized state measurements are available for designing the control signal. First, we discuss the stability of the system with quantized state measurements and then extend the results in the event-triggering framework.

Here, we also consider the sliding variable as $s = c^{\top}x$ and sliding manifold given by (1.14). We show that the control

$$u = -(c^{\top}B)^{-1}\left(c^{\top}Ax + K\operatorname{sign}s\right)$$

ensures the sliding mode in the system if $K > |c^{\top}B|d_0$. In this chapter, we consider the design of SMC with quantized state measurement. Following type of quantizer is considered in this work.

6.1.1 Quantizer

In this chapter, a quantizer with fixed sensitivity is considered for simplicity. A quantizer is a piecewise constant function $q : \mathbb{R} \rightarrow \mathscr{Q}$ that maps to a point in $\mathbb{R}$ a finite subset $\mathscr{Q}$ of $\mathbb{R}$. The finite subset $\mathscr{Q}$ partitions the whole state space into quantization regions and these regions need not be identical. Each quantizer is characterized by some level denoted by M and the sensitivity v. We call the quantizer is saturated when the quantized values do not belong to the union of M quantization regions. By sensitivity of the (uniform) quantizer, we mean that the maximum range of quantization region such that all signal are mapped to the same quantization region. We denote $q(z)$ as a quantized value of any given $z \in \mathbb{R}$ with $q(z) \in \mathscr{Q}$. Mathematically, the quantizer is represented as

$$q(z) = \begin{cases} Mv & \text{if } z \geq \left(M - \dfrac{1}{2}\right)v, \\[2mm] \left\lfloor \dfrac{z}{v} + \dfrac{1}{2} \right\rfloor v & \text{if } -\left(M - \dfrac{1}{2}\right)v \leq z < \left(M - \dfrac{1}{2}\right)v, \\[2mm] -Mv & \text{if } z < -\left(M - \dfrac{1}{2}\right)v \end{cases} \tag{6.1}$$

Such a quantizer is shown in Fig. 6.1 with saturation level M and sensitivity v. Here, $\lfloor z \rfloor$ denotes *floor* function that returns greatest integer smaller than or equal to z. In the case of vector, the quantizer is defined componentwise. So, for any $z \in \mathbb{R}^n$, the n quantizers can be denoted as $\left[q_1(z_1)\ q_2(z_2)\ \cdots\ q_n(z_n)\right]^{\top}$. It may be noted that each such quantizers may have different saturation levels and sensitivities given by $M_1, M_2, \ldots, M_n$ and $v_1, v_2, \ldots, v_n$, respectively. However, we assume a fixed saturation level and sensitivity for all n quantizers, i.e. $M_1 = \cdots = M_n =$

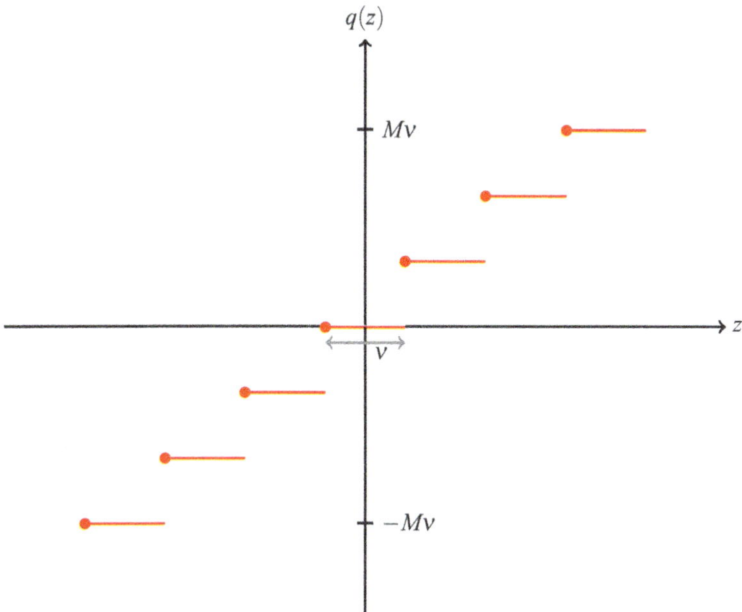

Fig. 6.1 A quantizer with sensitivity v and saturation level M

M and $v_1 = \cdots = v_n = v$. So, in the vector case, the quantizer is denoted as $\begin{bmatrix} q(z_1)\ q(z_2)\ \cdots\ q(z_n) \end{bmatrix}^{\top}$ where $q(\cdot)$ is with some fixed level and sensitivity. In this chapter, we write the quantizer $q(\cdot)$ for both scalar and vector case in respective contexts if there is no confusion arises.

Due to the quantization, some error is introduced in the system if the quantized values are used in the feedback control law. Define this error as $e_q = x - q(x)$. It can be seen that if the quantizer does not saturate, then we can have

$$\|e_q\| \leq v \frac{\sqrt{n}}{2}. \tag{6.2}$$

Here, the quantized control law is designed by replacing the states by their quantized values. So, SMC can be given as

$$u = -(c^{\top}B)^{-1}\left(c^{\top}Aq(x) + K\operatorname{sign}q(s)\right) \tag{6.3}$$

where $q(s) = c^{\top}q(x)$.

In the next section, we first give the stability of SMC with quantized measurements and then analyse the stability of the system in the event-triggering framework.

6.2   Design of Sliding Mode Control

If the control (6.3) is applied to the plant, the value of K must be chosen to guarantee that sliding trajectories reach in the vicinity of sliding surface in finite-time. So, we select K as

$$K \geq \|c\|\|A\|v\frac{\sqrt{n}}{2} + \left|c^\top B\right| d_0. \tag{6.4}$$

In the following, we state that with the choice of K given by (6.4), the convergence of sliding trajectory to the vicinity of sliding manifold is achieved.

Theorem 6.1 *Consider the system* (1.11) *and the control law* (6.3). *Then, if K is designed as per* (6.4), *the sliding trajectories are ultimately bounded by the region*

$$\left\{ x \in \mathbb{R}^n : |s| \leq \frac{K\|c\|v\sqrt{n}}{K - \|c\|\|A\|v\frac{\sqrt{n}}{2} - \left|c^\top B\right| d_0} \right\} \tag{6.5}$$

in the vicinity of sliding surface.

Proof Consider the Lyapunov function $V = \frac{1}{2}s^2$. Then, differentiating V along the system trajectory and using the control (6.3) yields

$$\begin{aligned}
\dot{V}(s(t)) &= s(t)\dot{s}(t) \\
&= s(t)\left(c^\top Ax(t) + c^\top Bu(t) + c^\top Bd(t)\right) \\
&= -s(t)K\,\mathrm{signq}(s(t)) + s(t)c^\top Ae_q(t) + s(t)c^\top Bd(t).
\end{aligned} \tag{6.6}$$

Recalling (6.2), we have $s(t) = c^\top x(t) = c^\top(q(x(t)) + e_q(t)) = q(s(t)) + c^\top e_q(t)$.

Using this, in the above, we obtain

$$\begin{aligned}
\dot{V}(s(t)) &= -K|q(s(t))| - Kc^\top e_q(t)\mathrm{signq}(s(t)) + s(t)c^\top Ae_q(t) + s(t)c^\top Bd(t) \\
&\leq -K|q(s(t))| + K\|c\|\|e_q(t)\| + |s(t)|\|c\|\|A\|\|e_q(t)\| + |s(t)|\left|c^\top B\right| d_0 \\
&\leq -K(|s(t)| - \|c\|\|e_q(t)\|) + K\|c\|\|e_q\| + |s(t)|\|c\|\|A\|\|e_q(t)\| + |s(t)|\left|c^\top B\right| d_0 \\
&= -K|s(t)| + 2K\|c\|\|e_q(t)\| + |s(t)|\|c\|\|A\|\|e_q(t)\| + |s(t)|\left|c^\top B\right| d_0 \\
&\leq -K|s(t)| + K\|c\|v\sqrt{n} + |s(t)|\|c\|\|A\|v\frac{\sqrt{n}}{2} + |s(t)|\left|c^\top B\right| d_0 \\
&\leq -|s(t)|\left(K - \|c\|\|A\|v\frac{\sqrt{n}}{2} - \left|c^\top B\right| d_0\right) + K\|c\|v\sqrt{n}.
\end{aligned} \tag{6.7}$$

Since K satisfies the condition (6.4), the ultimate bound of the system trajectories can be given as

$$\dot{V} \leq -\left(K - \|c\|\|A\|v\frac{\sqrt{n}}{2} - \|c^\top B\| d_0\right)$$

$$\times \left(|s(t)| - \frac{K\|c\|v\sqrt{n}}{K - \|c\|\|A\|v\frac{\sqrt{n}}{2} - \|c^\top B\| d_0}\right). \tag{6.8}$$

Thus, the ultimate bound of the sliding trajectories is given by (6.5). This completes the proof.

Remark 6.1 It can be seen from (6.5) that the band size is dependent on the sensitivity of quantizer and it can be reduced by selecting an appropriate v. In [11], a strategy is developed to vary the sensitivity of quantizer that achieves asymptotic stability. However, in the present chapter, such quantizer is not considered and it is assumed to be fixed one.

In the practical implementation, the control is updated in digital fashion through the digital processor. In this chapter, we emphasize the event-triggered implementation of control law when quantized measurements are transmitted to the controller end. In the previous chapter, the event-triggered SMC is discussed in detail. Here, we analyse the performance of the system with control law given by (6.3) in the event-triggered framework. So, the discrete quantized SMC can be given as

$$u(t) = -(c^\top B)^{-1}\left(c^\top Aq(x(t_i)) + K\operatorname{signq}(s(t_i))\right) \tag{6.9}$$

for all $t \in [t_i, t_{i+1})$. Before proceeding further, we define the following

$$s_q(t) = q(x(t_i)) - q(x(t)). \tag{6.10}$$

Similarly, $e(t) = x(t_i) - x(t)$ is considered as discrete error which is defined earlier. If the event generator is located at the controller end, then only quantized values are available and event detection may not be possible. So, in the following analysis, it is considered that the event generator is placed near the sensor end where actual state measurements are available.

6.3 Design of Event-Triggered Sliding Mode Control

In this section, the event-triggered implementation of SMC with quantized measurements is presented. Since the event is continuously monitored near the sensor end, the direct state measurements are available to the event-triggered block. So, we consider the event in this chapter as

$$\|e(t)\| \leq \sigma\alpha \tag{6.11}$$

where the design of the parameter α will be explained later and $\sigma \in (0, 1)$. At first, we establish the convergence of sliding trajectories to the vicnity of sliding manifold.

Theorem 6.2 *Consider the system* (1.11) *and the quantized control* (6.9). *Let* $\alpha_1 > 0$ *be given. Then the sliding trajectories are ultimately bounded within the region given by*

$$\left\{ x \in \mathbb{R}^n : |s| \leq \frac{2K \|A\|^{-1} \alpha_1 + K \|c\| v \sqrt{n}}{K - \|c\| \|A\| v \frac{\sqrt{n}}{2} - \alpha_1 - |c^\top B| d_0} \right\} \tag{6.12}$$

if the gain K is chosen as

$$K \geq \|c\| \|A\| v \frac{\sqrt{n}}{2} + \alpha_1 + |c^\top B| d_0 \tag{6.13}$$

where α_1 satisfies

$$\|c\| \|A\| \|s_q(t)\| \leq \alpha_1 \tag{6.14}$$

for all time.

Proof We first consider the time interval $t \in [t_i, t_{i+1})$. Select the Lyapunov function $V(s) = \frac{1}{2} s^2$. Now, differentiating V with respect to time along the system trajectories, we get

$$\begin{aligned} \dot{V}(s(t)) &= s(t)\dot{s}(t) \\ &= s(t)\left(c^\top A x(t) + c^\top B u(t) + c^\top B d(t)\right). \end{aligned} \tag{6.15}$$

Substituting (6.9) for $u(t)$ in (6.15),

$$\begin{aligned} \dot{V}(s(t)) &= s(t)\left(c^\top A x(t) - c^\top A q(x(t_i)) - K\,\mathrm{signq}(s(t_i)) + c^\top B d(t)\right) \\ &= s(t)\big(c^\top A x(t) - c^\top A q(x(t)) + c^\top A q(x(t)) - c^\top A q(x(t_i)) - K\,\mathrm{signq}(s(t_i)) \\ &\quad + c^\top B d(t)\big) \\ &= s(t)\left(c^\top A e_q(t) - c^\top A s_q(t) - K\,\mathrm{signq}(s(t_i)) + c^\top B d(t)\right). \end{aligned} \tag{6.16}$$

Using the relation $x(t) = q(x(t)) + e_q(t)$, it can be easily shown that $s(t) = q(s(t_i)) - c^\top s_q(t) + c^\top e_q(t)$. With the help of these equation, the above relation can be reduced to

$$\dot{V}(s(t)) = -(q(s(t_i)) - c^\top s_q(t) + c^\top e_q(t)) K \operatorname{sign} q(s(t_i)) + s(t) c^\top A e_q(t)$$
$$- s(t) c^\top A s_q(t) + s(t) c^\top B d(t)$$
$$\leq -K |q(s(t_i))| + K \|c\| \|s_q(t)\| + K \|c\| \|e_q(t)\| + |s(t)| \|c\| \|A\| \|e_q(t)\|$$
$$+ |s(t)| \|c\| \|A\| \|s_q(t)\| + |s(t)| \left| c^\top B \right| d_0. \tag{6.17}$$

We know that $q(x(t_i)) = s_q(t) + x(t) - e_q(t)$, so we can write $s(t) = q(s(t_i)) + c^\top e_q(t) - c^\top s_q(t)$. Using triangle inequality, we obtain $|s(t)| \leq |q(s(t_i))| + \|c\| \|e_q(t)\| + \|c\| \|s_q(t)\|$. Now using these in (6.17), we obtain

$$\dot{V}(s(t)) \leq -K(|s(t)| - \|c\| \|e_q(t)\| - \|c\| \|s_q(t)\|) + K \|c\| \|s_q(t)\| + K \|c\| \|e_q(t)\|$$
$$+ |s(t)| \|c\| \|A\| \|e_q\| + |s(t)| \|c\| \|A\| \|s_q\| + |s(t)| \left| c^\top B \right| d_0$$
$$= -K |s(t)| + 2K \|c\| \|s_q(t)\| + 2K \|c\| \|e_q(t)\| + |s(t)| \|c\| \|A\| \|e_q(t)\|$$
$$+ |s(t)| \|c\| \|A\| \|s_q\| + |s(t)| \left| c^\top B \right| d_0$$
$$\leq -K |s(t)| + 2K \|A\|^{-1} \alpha_1 + K \|c\| v \sqrt{n} + |s(t)| \|c\| \|A\| v \frac{\sqrt{n}}{2}$$
$$+ |s(t)| \alpha_1 + |s(t)| \left| c^\top B \right| d_0$$
$$= -|s(t)| \left(K - \|c\| \|A\| v \frac{\sqrt{n}}{2} - \alpha_1 - \left| c^\top B \right| d_0 \right) + 2K \|A\|^{-1} \alpha_1 + K \|c\| v \sqrt{n}$$
$$= -\left(K - \|c\| \|A\| v \frac{\sqrt{n}}{2} - \alpha_1 - \left| c^\top B \right| d_0 \right)$$
$$\times \left(|s(t)| - \frac{2K \|A\|^{-1} \alpha_1 + K \|c\| v \sqrt{n}}{K - \|c\| \|A\| v \frac{\sqrt{n}}{2} - \alpha_1 - \left| c^\top B \right| d_0} \right). \tag{6.18}$$

Thus, the ultimate bound is given by (6.12) and outside this band, the trajectories are attracted towards the sliding surface. The size of the band can be reduced as per the desired value by suitably designing the different parameters. This completes the proof.

6.3.1 Design of Event-Triggering Rule

In the next, we discuss how the condition (6.14) is satisfied for all time. As per definition, it is known that

$$\|s_q(t)\| = \|q(x(t_i)) - q(x(t))\|$$
$$= \|q(x(t_i)) - x(t_i) + x(t_i) - q(x(t))\|$$
$$\leq \|q(x(t_i)) - x(t_i)\| + \|x(t_i) - q(x(t))\|$$
$$\leq v \frac{\sqrt{n}}{2} + \|x(t_i) - x(t) + x(t) - q(x(t))\|$$

$$\leq \nu\frac{\sqrt{n}}{2} + \|e(t)\| + \nu\frac{\sqrt{n}}{2}. \tag{6.19}$$

Using the relation (6.11) in the above expression, we obtain the upper bound of the error due to quantization as

$$\|s_q(t)\| \leq \nu\sqrt{n} + \sigma\alpha. \tag{6.20}$$

Define $\alpha_1 := (\nu\sqrt{n} + \sigma\alpha)\|c\|\|A\|$. Hence, by restricting the condition (6.11), we can guarantee (6.14) holds for all time. It may be noted that given some $\alpha_1 > 0$, the value of α can be obtained from the above relation. It is be noted that the condition $\|e(t)\| \leq \sigma\alpha$ requires direct state measurements, so it is located at the sensor ends where the actual values of state are available. Indeed, this ensure that no Zeno phenomenon occurs in the event-triggering implementation. Hence, the condition (6.11) is used to design the triggering condiion and the triggering instants can be found as

$$t_{i+1} = \inf\{t > t_i : \|e(t)\| \geq \sigma\alpha\}. \tag{6.21}$$

Once the sliding trajectories become bounded for all time, the state trajectories also remain bounded. It can be easily shown as follows. When the system is in sliding phase

$$|s(t)| \leq \alpha^\star$$

where $\alpha^\star$ denotes the ultimate bound of sliding trajectories given by (6.12). It follows immediately $x_2(t) = -c_1^\top x_1(t) + s(t)$. Then, the reduced dynamics can be written as

$$\dot{x}_1(t) = (A_{11} - A_{12}c_1^\top)x_1(t) + A_{12}s(t). \tag{6.22}$$

The following proposition gives the ultimate bound of the system state trajectories when trajectories are in the vicinity of sliding manifold.

Proposition 6.1 *Consider the system (6.22). Then the trajectories remain ultimately bounded in the region given by*

$$\left\{ x_1 \in \mathbb{R}^{n-1} : \|x_1\| \leq \frac{2\alpha^\star \|A_{12}^\top P\|}{\lambda_{\min}\{Q\}} \right\} \tag{6.23}$$

where P and Q are symmetric positive-definite matrices such that

$$(A_{11} - A_{12}c_1^\top)^\top P + P(A_{11} - A_{12}c_1^\top) = -Q.$$

Proof Proof directly follows from stability of LTI systems.

Now we will show that the event-triggering condition (6.11) ensures no accumulation of triggering sequences $\{t_i\}_{i=0}^{\infty}$ generated by (6.21). This result is similar to that of presented in Theorem 2.2.

Theorem 6.3 *Consider the system* (1.11) *and triggering condition* (6.21). *Let the control law* (6.9) *brings the sliding mode in the system by executing the event* (6.21) *for all* $t > t_i$ *for the increasing time sequence* $\{t_i\}_{i\in\mathbb{N}_0}$, *i.e.* $t_0 < t_1 < t_2 < \cdots$. *Then, the inter-execution time* $T_i = t_{i+1} - t_i$ *satisfies*

$$T_i \geq \frac{1}{\|A\|} \ln\left(1 + \frac{\sigma\alpha\|A\|}{\rho_Q(\|x(t_i)\|) + \beta}\right) \tag{6.24}$$

where the real-valued function $\rho_Q(\|x(t_i)\|) : \mathbb{R}_{\geq 0} \mapsto \mathbb{R}_{\geq 0}$ *and* β *are defined as*

$$\rho_Q(\|x(t_i)\|) := \left\| Ax(t_i) - B(c^{\mathsf{T}}B)^{-1}c^{\mathsf{T}}Aq(x(t_i)) \right\| \tag{6.25}$$

and

$$\beta := \left\| B(c^{\mathsf{T}}B)^{-1}K \right\| + \|B\| d_0. \tag{6.26}$$

Proof Consider the set $\Gamma = \{t \in [t_i, t_{i+1}) : \|e(t)\| = 0\}$. Now, taking the time derivative of $\|e(t)\|$ for all time $t \in [t_i, t_{i+1})\backslash\Gamma$,

$$
\begin{aligned}
\frac{\mathrm{d}}{\mathrm{d}t}\|e(t)\| &\leq \left\|\frac{\mathrm{d}}{\mathrm{d}t}e(t)\right\| = \left\|\frac{\mathrm{d}}{\mathrm{d}t}x(t)\right\| \\
&= \left\| Ax(t) - B(c^{\mathsf{T}}B)^{-1}c^{\mathsf{T}}Aq(x(t_i)) - B(c^{\mathsf{T}}B)^{-1}K\operatorname{signq}(s(t_i)) + Bd(t) \right\| \\
&= \left\| Ax(t_i) - Ae(t) - B(c^{\mathsf{T}}B)^{-1}c^{\mathsf{T}}Aq(x(t_i)) - B(c^{\mathsf{T}}B)^{-1}K\operatorname{signq}(s(t_i)) + Bd(t) \right\| \\
&\leq \|A\|\|e(t)\| + \left\| Ax(t_i) - B(c^{\mathsf{T}}B)^{-1}c^{\mathsf{T}}Aq(x(t_i)) \right\| + \|B(c^{\mathsf{T}}B)^{-1}K\| + \|B\|d_0 \\
&= \|A\|\|e(t)\| + \rho_Q(\|x(t_i)\|) + \beta. \tag{6.27}
\end{aligned}
$$

The solution to the above differential inequality is obtained using comparison Lemma with $\|e(t_i)\| = 0$ as initial condition,

$$\|e(t)\| \leq \frac{\rho_Q(\|x(t_i)\|) + \beta}{\|A\|}\left(e^{\|A\|(t-t_i)} - 1\right) \tag{6.28}$$

for all time $t \in [t_i, t_{i+1})$. At the triggering instant, $\|e(t_{i+1})\| = \sigma\alpha$, so

$$\sigma\alpha = \|e(t_{i+1})\| \leq \frac{\rho_Q(\|x(t_i)\|) + \beta}{\|A\|}\left(e^{\|A\|T_i} - 1\right). \tag{6.29}$$

The rearrangement of relation (6.29) gives the relation (6.24). It can be seen that for any finite value of state, the lower bound of inter-event time is strictly greater than zero. Thus, a positive lower bound for the inter-event time is ensured. This completes the proof. $\square$

Remark 6.2 The event-triggering rule (6.21) in quantized SMC is similar to (2.12) as the direct state measurements are used by the triggering mechanism. But, the control signal requires quantized values only in the quantized SMC. It might happen that no control signal is updated in the quantized event-triggering case even if the triggering instant is generated. However, this is not with the case that appears in [18].

The event parameter α must be chosen sufficiently large such that the quantized values are different. Note that the consecutive quantized values differ by v. Now select the value for α as

$$\alpha > v. \tag{6.30}$$

This choice of event parameter will always ensure that there is always a change in the quantized values whenever an event is triggered.

6.4   Simulation Results

In this section, we present numerical simulation pertaining to the theoretical analysis given in the chapter. We consider system as

$$\dot{x}(t) = \begin{bmatrix} 0 & 1 \\ 4 & 5 \end{bmatrix} x(t) + \begin{bmatrix} 0 \\ 1 \end{bmatrix} (u(t) + 0.5\sin(10t)).$$

The sliding surface is designed as $s(t) = \begin{bmatrix} 0.5 & 1 \end{bmatrix} x(t)$. The following parameters are chosen as $K = 1.8276$, $\alpha = 0.0693$, $\sigma = 0.85$. The initial condition is chosen as $x_0 = \begin{bmatrix} 1 & 1 \end{bmatrix}$. The sensitivity and saturation level of quantizer are chosen as $v = 0.1$ and $M = 50$. We assume that the quantizer does not saturate in the working range of the system. The control input for the system is given as

$$u(t) = -\left(\begin{bmatrix} 4 & 5.5 \end{bmatrix} \begin{bmatrix} q(x_1(t_i)) \\ q(x_2(t_i)) \end{bmatrix} + K\,\mathrm{sign}q(s(t_i)) \right)$$

for all $t \in [t_i, t_{i+1})$.

Figure 6.2 shows the plots of state trajectory and sampling interval. It can be seen that states of the system converge and remain ultimately bounded in the vicinity of sliding manifold as shown in Fig. 6.2a. In this case, the ultimate bound of state trajectory depends on both quantizer and event parameters. It may also be noted that

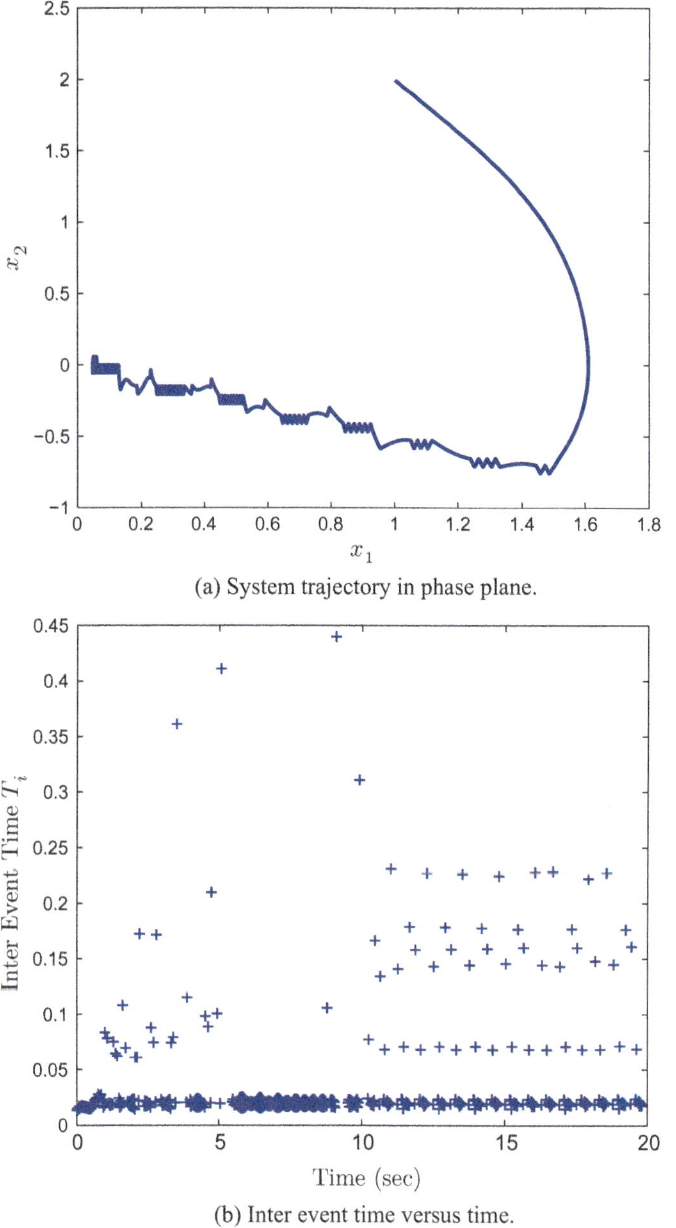

(a) System trajectory in phase plane.

(b) Inter event time versus time.

Fig. 6.2 Performance of event-triggered SMC for LTI systems with quantized measurements

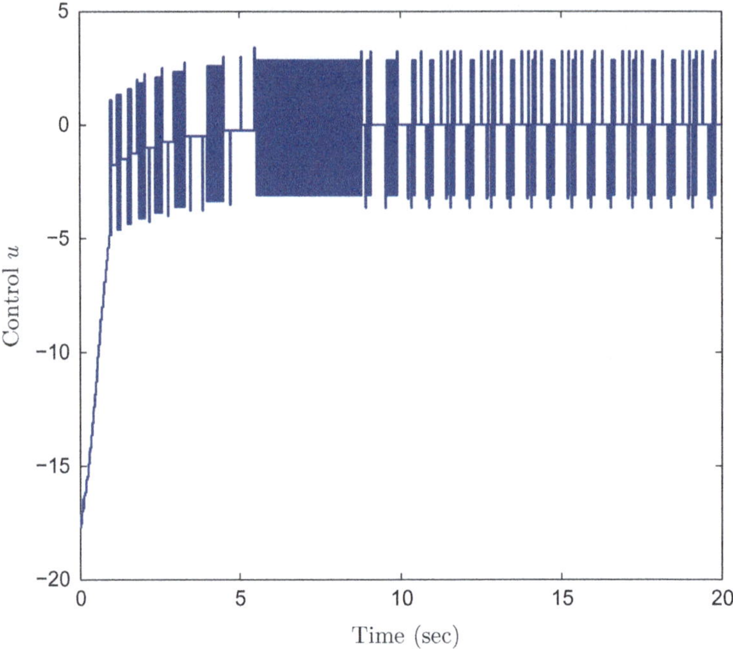

Fig. 6.3 Event-triggered quantized SMC signal for LTI systems

the ultimate bound decreases with decrease in the sensitivity of the quantizer. Further, the Zeno execution of event-triggering condition is avoided as shown in Fig. 6.2b. The control signal is plotted in Fig. 6.3.

6.5 Summary

This chapter presents the event-triggered SMC with quantized measurement. First, the quantized SMC is presented for LTI systems. Here, the stability of sliding motion and system behaviour during sliding mode are given in detail. In the event-triggered SMC with quantized measurement, the sufficient condition is given for the stability of closed loop system. Finally, the numerical example is considered to illustrate the design of quantized SMC with event-triggering strategy.

6.6 Notes and References

The design of SMC is discussed in detail in [1–7] and reference therein. Stability of systems with quantized measurements is studied in [8, 9], and quantized SMC in [10–12]. For the preliminaries idea on the event-triggered control, refer to [13–17], and for event-triggered SMC, the readers may see [18–23]. In [20], the quantized event-triggered SMC is presented as a preliminary result.

References

1. V.I. Utkin, Variable structure systems with sliding modes. IEEE Trans. Autom. Control **22**(2), 212–222 (1977)
2. V.I. Utkin, *Sliding Modes and Their Applications in Variable Structure Systems*, Translated from the Russian by A. Parnakh (MIR Publishers, Moscow, 1978)
3. D. Draženović, The invariance conditions in variable structure systems. Automatica **5**(3), 287–295 (1969)
4. V.I. Utkin, *Sliding Modes in Control and Optimization* (Springer, New York, 1992)
5. V.I. Utkin, J. Gulnder, J. Shi, *Sliding Mode Control in Electromechanical Systems* (CRC Press, Taylor and Francis Group, 1999)
6. A.F. Filippov, *Differential Equations With Discontinuous Right-Hand Sides* (Kluwer Academic Publishers, Dordrecht, The Netherlands, 1988)
7. C. Edwards, S.K. Spurgeon, *Sliding Mode Control: Theory and Applications* (CRC Press, Taylor and Francis Group, 1998)
8. R.G. Brockett, D. Liberzon, Quantized feedback stabilization of linear systems. IEEE Trans. Autom. Control **45**(7), 1279–1289 (2000)
9. N. Elia, S.K. Mitter, Stabilizing linear systems with minimal information. IEEE Trans. Autom. Control **46**(9), 1384–1400 (2001)
10. M.L. Corradini, G. Orlando, Robust quantized feedback stabilization of linear systems. Automatica **44**(9), 2458–2462 (2008)
11. B.-C. Zheng, G.-H. Yang, Robust quantized feedback stabilization of linear systems based on sliding mode control. Optimal Control Appl. Methods **34**(4), 458–471 (2013)
12. B.-C. Zheng, G.-H. Yang, Quantized output feedback stabilization of uncertain systems with input nonlinearities via sliding mode control. Int. J. Robust Nonlinear Control **24**(2), 228–246 (2014)
13. K.-E. Årźen, A simple event-based PID controller, in *Proceedings of 14th IFAC World Congrress*, Beijing, China (1999), pp. 423–428
14. K.J. Åström, B. Bernhardsson, Comparison of Riemann and Lebesgue sampling for first order stochastic systems, in *Proceedings of 41st IEEE Conference on Decision and Control*, Las Vegas, USA (2002), pp. 2011–2016
15. P. Tabuada, Event-triggered real-time scheduling of stabilizing control tasks. IEEE Trans. Autom. Control **52**(9), 1680–1685 (2007)
16. Y.-K. Xu, X.-R. Cao, Lebesgue-sampling-based optimal control problems with time aggregation. IEEE Trans. Autom. Control **56**(5), 1097–1109 (2011)
17. W.P.M.H. Heemels, K.H. Johansson, P. Tabuada, An introduction to event-triggered and self-triggered control, in *Proceedings of 51st IEEE Conference of Decision and Control*, Hawai, USA (2010), pp. 3270–3285
18. A.K. Behera, B. Bandyopadhyay, Event based robust stabilization of linear systems, in *Proceedings of 40th Annual Conference of the IEEE Industrial Electronics Society*, Dallas, USA (2014), pp. 133–138

19. A. Ferrara, G.P. Incremona, L. Magini, Model-based event-triggered robust MPC/ISM, in *Proceedings of 13th European Control Conference*, Strausbourg, France (2014), pp. 2931–2936
20. A.K. Behera, B. Bandyopadhyay, Event based sliding mode control with quantized measurement, in *Proceedings of International Workshop on Recent Advances Sliding Modes*, Istanbul, Turkey (2015), pp. 1–6
21. A.K. Behera, B. Bandyopadhyay, Self-triggering-based sliding-mode control for linear systems. IET Control Theory Appl. **9**(17), 2541–2547 (2015)
22. A.K. Behera, B. Bandyopadhyay, Event-triggered sliding mode control for a class of nonlinear systems. Int. J. Control **89**(9), 1916–1931 (2016)
23. A.K. Behera, B. Bandyopadhyay, Robust sliding mode control: an event-triggering approach. IEEE Trans. Circuits Syst. II Express Briefs **64**(2), 146–150 (2017)

Index

© Springer International Publishing AG, part of Springer Nature 2018   127
B. Bandyopadhyay and A. K. Behera, *Event-Triggered Sliding Mode Control*, Studies
in Systems, Decision and Control 139, https://doi.org/10.1007/978-3-319-74219-9

The manufacturer's authorised representative in the EU is Springer
Nature Customer Service Centre GmbH, Europaplatz 3, 69115 Heidelberg,
Germany. If you have any concerns regarding our products, please
contact ProductSafety@springernature.com

Printed and bound by CPI Group (UK) Ltd, Croydon, CR0 4YY

28/11/2025

02007724-0002